EBURY PRESS

GHOSTS ON PEEPAL TREES

Swami Prem Parivartan, known to millions as Peepal Baba, is an environmentalist who, along with his team, has planted and preserved over 26 million trees in over 226 districts across twenty-one states in India. He started his work with trees at the age of eleven. He founded Give Me Trees Trust, an NGO for ecological and habitat restoration in 2011.

Deeply influenced by the teachings of Osho, he joined the Neo-Sannyas Movement, where he was given the name 'Swami Prem Parivartan'. He reaches out to millions of followers through his social media platforms, which feature his talks on nature conservation.

Ghosts *on* Peepal Trees

My Journey from Folktales to Forests

Peepal Baba

EBURY PRESS
An imprint of Penguin Random House

EBURY PRESS

Ebury Press is an imprint of the Penguin Random House group of companies
whose addresses can be found at global.penguinrandomhouse.com

Published by Penguin Random House India Pvt. Ltd
4th Floor, Capital Tower 1, MG Road,
Gurugram 122 002, Haryana, India

Penguin Random House India

First published in Ebury Press by Penguin Random House India 2026

10 9 8 7 6 5 4 3 2 1

ISBN 9780143476702

Typeset in Minion Pro by MAP Systems, Bengaluru, India

www.penguin.co.in

To my grandmother, Pushpa Amar, and my first teacher, Mrs Williams, who nurtured both my roots and my wings

To the peepal trees I've cared for and to the ghosts who have been my unseen family, thank you for your silent love and protection

Contents

'Most memoir writers will tell you that the hardest part of writing a memoir isn't what to include, but what to leave out.'

—Kathleen Flinn

Author's Note

This book is a tribute, not to the numbers but to the souls behind them. To the countless volunteers, children, farmers, monks, students and wanderers who joined hands in this holy exercise of adding green to the earth. From the one who planted a single tree to those who planted thousands, this is my salute to the collective spirit of community work.

I am not here to list the names of people I met, nor to recount the kilometres of roads or every species we covered. This is not a research-based, scholarly treatise on the technicalities of trees and the environment. Taxonomy has never come close to me, nor has modern tinkering with genes, mutations or tissue cultures. I am a student of literature, and I have always seen poetry in the world of plants. I am here to share the inner movement, the rhythm that began in one heart and rippled through many.

The seed of every movement is belief. It begins with a whisper, the wish to do something good, and grows into a forest only because others feel the same heartbeat. That is the story I wish to tell.

This thought of a book began in 2012, and it took over a decade and a road accident in November 2024 to finally find the time and the stillness to wrap it up. The ghosts took a long time to settle in. I wrote in fits and starts, pushing myself to gather words like fallen leaves. I was clear about one thing: my experiences had come to me the hard way, and it was only right to share them before it was too late. What I went through while writing these pages is nothing short of creating a new me. So many experiences that I had once passed over crystallized and fell right in front of me.

Foremost among them is the realization that nature is far more intelligent than humans. Our only true purpose is to celebrate what evolution—4.5 billion years of silent genius—has handed down to us. Yet, in our brief seventy or eighty years, we insist on rearranging everything. The understanding of science should lead to awe, not arrogance. Instead, we tinker, theorize and forget that we are transient guests on this blue planet.

After working on the ground for nearly five decades, restoring vegetation across 2,70,000 hectares, planting twenty-five million trees and as many shrubs, I confess: I have hardly made a dent. But I tried. That is what matters.

The chapters in this book are not merely stories. They are symbols of our Indian psyche, woven through our culture, scriptures, myths and memories. Some beliefs have grown into institutions of their own, sacred and untouchable. They are part of our collective DNA.

The Peepal tree fascinated me from childhood. I have spent forty-eight years planting, preserving and promoting this keystone species. This book is my way of sharing that long companionship—bitter, sweet and sacred. Welcome to the kingdom of peepal trees.

Let's talk about the simple things that shape our lives. Let's climb down from our human pedestal and see the world as it is: mysterious, miraculous, alive. Let's talk about our teachers, our grandmothers, our friends and the neighbourhoods that raised us.

This is about devotion. I just want to show how easy it is to fall in love with greens and biodiversity, to take small, beautiful steps that matter. I have always written for myself. I enjoyed reading my own diaries and revisiting memories. Only when my editors nudged me did I decide to share my private joys with the world. This is a memoir of exploration, of ghosts and angels, of trees and teachers, of the quiet miracle of existence. It's about the soil, the saplings, the smiling faces that kept me going. It's about a life that happened on its own, like a forest. I share this so you may find your own beginning in these stories. I hope you see what I saw—the poetry, the laughter, the small acts of grace that make a life worth living.

While writing this memoir, I discovered myself. I enjoyed the process. I hope you enjoy the journey.

With love,
Peepal Baba

Chapter 1

Early Years

I was born in Chandigarh in 1966, in the clean geometry of a city still finding its rhythm. My father was then a young army doctor, holding the rank of captain. When I opened my eyes at PGI Hospital—one of the greenest corners of India—my father was posted on the border in the Lahore sector. He came to see me the next day, still in his battle fatigues, his boots carrying the scent of the field. My mother, meanwhile, was completing her postgraduation in Hindi Literature at Punjab University.

For the first few days of my life, I was raised in the girls' hostel, surrounded by laughter, saris hung out to dry. I had no role to play in that decision, of course, but it's a story that always makes me smile: my first cradle in a women's hostel, rocked by conversations about poetry and politics.

Old family photos laid out on a table

During my first four years on this planet, I lived and travelled through four cities. My father had seen me as a newborn wrapped in hospital linen and then returned to the border. Orders followed; our small family was posted to Dalhousie in Himachal Pradesh. As soon as my mother completed her studies, we packed our bags—those black aluminium trunks with white stencilled ranks, the tin boxes of crockery, the army bedding rolls—and moved from one beautiful city to another. Cleanest and greenest—that seemed to be God's pattern for me.

Dalhousie was a jewel set in deodar and mist. My father was allotted a sprawling bungalow overlooking a valley in the cantonment area. In those days, few ventured up to the hill station; the road from Pathankot was a ribbon of broken stones that twisted through pine forests and clouds. My mother later told me that she would walk me in a woollen wrap under those tall deodars, pointing to monkeys and the glitter of mica on the road. Perhaps that is why the smell of conifers follows me everywhere. The resin of those trees entered my breath before language did.

The postings changed again; first Calcutta (now Kolkata), and then Dehradun, when my father was transferred to the local military hospital from Calcutta in 1970. It was as if my early childhood had wheels. I have no memory of walls that didn't move, of trunks that weren't packed. My earliest images are of trains, railway platforms and the metallic smell of iron tracks warming under the afternoon sun. I remember movement itself: the sound of a whistle, the jolt of the carriage, my mother's hand steadying a milk bottle, my father's cap hanging from a hook, the sky framed by the window bars.

These cities were not merely places; they were lessons in light and leaf. I kept returning to them in later years, as if my roots had scattered themselves along those routes. But it was Dehradun where we finally camped for three full years, three generous years that quietly set the compass of my childhood and shaped everything I would later become. My school, my teachers, my walks with Granny, the hills, the trees and our pilgrimages to Mussoorie, Rishikesh, Haridwar, Uttarkashi, Badrinath and Kedarnath—all these are still etched in the inner bark of my memory.

My mother noticed something early: I cried indoors but grew quiet under open skies. That pattern never left me. Even decades later, during the pandemic lockdowns, I could not survive within walls. I watered plants, felt the sun on my back, spoke to the wind as if it were an old friend. Nature did not design us for confinement. It built open skies and called it home. Dehradun in the early seventies was a slow-beating heart of hills and rain. It was a city of teachers and soldiers, of bicycle bells and shaded roads. The air itself seemed educated. My father, now a major, served in the paediatrics department of the Military Hospital. The 1971 war that created Bangladesh raged across the news, but our small world remained an island of calm.

Cambrian Hall School became my first formal classroom. Before that, I had done a year each in Calcutta, Dalhousie and Chandigarh, a dizzying curriculum for a little boy. The teachers at Cambrian complained that I was too restless, too curious. I was often punished and sent out of class, which suited me perfectly. The outdoors was my real syllabus. The banyan by the gate, the sparrow nests under the eaves, the puddles that reflected the sky, those were my chapters. The teachers grew angrier when they discovered I was the happiest in punishment.

I remember one afternoon on the rooftop of our house, a sandwich in hand, when an eagle swooped down and snatched it away. For most, that would have been a shock; for me, it was a revelation. I began to read all I could about raptors, eagles, kites, hawks and vultures. That single act of theft had opened a lifelong friendship with the sky.

Years passed, postings changed, but Dehradun remained the city of my heart. Even now, when I think of belonging, it is the Doon valley that rises before my eyes, the smell of rain on khaki soil, the silver line of the Ganga, the silhouettes of the sal forests at dusk. There I learned patience, observation and the quiet rhythm of living things.

In June 2012, many years later, I met Ruskin Bond at Cambridge Book Depot on Mall Road, Mussoorie, the same hills that had cradled my childhood. We spoke briefly about sacred trees. He smiled and wrote on the book I bought that evening: '*To dear Peepal Baba.*' That inscription felt like a benediction from the mountains themselves. The next morning, I sat at Char Dukaan near St Patrick's Church in Landour, and on the back of a tea-stained napkin, I wrote the first four sentences of what would one day become this book.

It was then I realized that the ecosystem had been preparing me all along, from the hostel cradle in Chandigarh to the compost pits of Dehradun. Early years never leave you; they dissolve into your bloodstream. Chandigarh, Dalhousie, Dehradun—these were not just places I lived in; they were teachers wearing different landscapes. When I look back, I see how quietly the world was tuning me. My father, with his discipline and duty; my mother, with her art and patience. Nature with its beauty and humility. And Granny, with her wisdom and her conversations with ghosts.

This understanding did not come all at once; it had begun much earlier.

Chapter 2

Granny Takes Over

Granny came to help my mother when I was born, but in truth, she came to command the household. A watchtower in a white Punjabi salwar kameez, she would stay with us for the next twenty years, through Dalhousie, Kolkata and Dehradun, until she finally moved to Allahabad (now Prayagraj) to live with her youngest son. Twenty years in her orbit were enough to lay the foundations of my character. As people say, the first twenty decide the next sixty. From my birth to my graduation (1966–86), I found myself in her presence.

Granny was, by herself, a thriving ecosystem. When Granny entered a home, she didn't just take over the kitchen; she took over everything she touched. Pots obeyed her, doors stayed ajar for the breeze she approved and even the walls seemed to grow warmer in her presence. She had a way of settling the air, of turning ordinary rooms into breathing spaces. Looking back, I realize how complete she was: self-contained, inventive, infinitely patient with the world, yet impatient with disorder. She fascinated me not because she preached, but because she practised. At six, she taught me to make compost from kitchen leftovers. 'Waste nothing,' she'd say, handing me a bowl of vegetable peels. 'Even what you throw away has some shame; let it find new life.'

Every morning, it was my job to carry the kitchen scraps down to the backyard where two neat pits waited like silent mouths. I'd lift the lid, a tin sheet weighed down by a stone, and empty the bowl, watching the slow alchemy begin. To me, it was magic; to her, it was routine. From refuse, she created renewal. That patch of soil was my first schooling in the environment. There I learned that the earth listens, that water has moods, that a seed is a promise that never forgets.

'Keep the soil moist, not drowned,' Granny would warn, tapping the edge of the watering can.

'Yes, Amma,' I'd answer, trying to imitate her sternness.

'If you listen, the plants will tell you what they need.'

And indeed, they did. The beans sulked when overwatered, the spinach shimmered when pleased. I learned to watch their leaves for answers.

Our house in Dehradun backed onto a litchi orchard owned by a Nepali family, rumoured descendants of the royal line. In May, when the litchis ripened, their perfume filled the evenings. I'd climb the low branches, scrape my knees and come home sticky-fingered and proud. Granny would laugh. 'Sweetness demands a little pain,' she'd murmur, dabbing antiseptic on the cuts. Those words became a proverb in my life.

Our walks became lessons, too. Every Sunday, we walked to the bazar—Paltan Bazar, Chakrata Road, Rajpur Road, sometimes all the way to the foothills of Mussoorie. Granny greeted every vendor by name. She knew who sold the sweetest guavas, which sadhu had turned into a gardener and which rickshaw-puller was fasting that week. To me, she seemed to carry an invisible map of kindness. She stopped to speak to stray dogs, to bless children, to drop a handful of grain near temples. 'Everyone must eat,' she'd say, and that included sparrows.

It was she who introduced me to the forests of Uttarakhand, Corbett, Rajaji, Haridwar, Rishikesh, Narendra Nagar, Tehri, Uttarkashi, Nainital and Almora. We travelled in old jeeps and army buses, staying in rest houses that smelled of kerosene and monsoon. I remember bathing in the Ganga at Haridwar when its waters were still crystal clear, and walking across Lakshman Jhula when it was a bridge of silence. Saints, vendors, gardeners, ghosts—Granny knew them all. She belonged to every landscape.

Those were the days of earthen pots beneath peepal trees, of sadhus who told stories instead of sermons, of lanes where one could hear one's own footsteps. Granny seemed to be on first-name terms with every God. 'They listen better when you whisper,' she'd say, lighting a lamp under the peepal.

She taught me that roots can travel, that faith can compost itself into growth and that under every peepal tree, a ghost of wisdom waits to whisper to those who listen.

Granny's jurisdiction extended beyond human affairs: to trees, ghosts, fish and kitchen gardens. Her reign was total, and her subjects were living, breathing, growing things—soil, saplings, sunlight. By the time we reached Pune, my destiny was well under her management.

At the time, I did not realize it, but these small routines were shaping everything that would follow.

Chapter 3

Learning of Ghosts in Trees

When my father was posted from Dehradun to Poona (now Pune), we were allotted a huge bungalow on Range Hills Road, in one of the satellite military stations outside the city, which was widely considered the cultural and literary capital of Maharashtra.

Range Hills Road was the kind of road that looked like it had been drawn by a patient hand, one clean, elegant line through the quiet suburb of Khadki, just outside the city of Pune. In those days, the city was called Poona. It began at a cluster of rocky brown hills and ended where the railway tracks hummed like distant hymns. The road was barely a kilometre long, lined with sprawling army bungalows, their gardens swelling with bougainvillaea and nostalgia. Every house had its own little kingdom of life, lawns clipped to military precision, rows of crotons and a few mango trees that had seen better centuries.

The evening breeze here had a uniform of its own; it blew on time, never too strong, never too weak, as if saluting the passing hour. From our veranda at Bungalow T-23, the hills caught the last orange blush of sunset, and the trains in the distance announced the hour like temple bells.

Our bungalow sat on two acres of generous land, a British-era relic with wide verandas, tall windows, creaking floors and the faint perfume of old wood and kerosene lamps. It had eight rooms, two sets of servant quarters, a fish pond and trees that could have told you stories of soldiers who'd left for wars long forgotten. The day after we moved in, Granny inspected the grounds like an officer on duty. She walked the entire property, stick in hand, making notes aloud:

'Three neem trees, four mangoes, two peepal, one banyan. The fish pond needs cleaning. The soil is sandy. We'll need compost. Kitchen waste, Munna, not to be thrown, that's gold.'

The peepal tree near the main gate was already grand—decades old, with roots that clutched the earth like a secret. It stood tall, dignified, a few steps from the main gate, its heart-shaped leaves quivering in perpetual conversation with the wind. Beside it stood a banyan, older and wiser, its aerial roots cascading like the beard of an ancient sage. It was a forest unto itself, branches, roots and birds forming an entire world that barely noticed us humans below. Together, they presided over the compound like twin guardians, one watching the day, the other ruling the night, and both ruling my imagination.

Granny said both were sacred. The gardener said both were haunted. I believed them both.

People from the neighbourhood often stopped at our gate to offer water, flowers or prayers to these trees. Some tied red threads around their trunks. Some whispered wishes. Some left quietly, heads bowed. The trees received it all with patience, a patience I have spent my whole life trying to learn.

By the end of the week, a vegetable garden had materialized behind the house—tomatoes, spinach, gourds, chillies, brinjals, all planted in neat lines like soldiers in a parade. The front lawn was hers, too. She grew marigolds for worship, roses for pride, hibiscus for shoe polish and jasmine for dreams. She had a rule for everything: 'Never water in harsh sun. Talk to plants before you trim them. Praise the gardener: scold the goats.'

The gardener, Maali Kaka, as we would call him, a quiet man from Satara, followed her orders like scripture. And I, her only grandchild in the household, was conscripted as her assistant. My duty was to carry the watering can, fetch earthworms and stay alert for lessons. It was in that garden that Granny gave me my first title. 'You are now a *gardener*,' she announced, handing me a trowel as if it were a royal sword. I was eleven years old, barefoot, and slightly confused. But she said it with such authority that it became true. Every morning began with the smell of wet earth and boiled milk. Granny would hum old Hindi songs while cutting spinach, and I would sit near her, counting the number of saplings that had survived the night. She made gardening sound like a sacred ritual.

'Plant with faith,' she'd say. 'Water with patience. Talk to them like your children.' By noon, she'd move to the kitchen, where she reigned with the same precision she used on her flowerbeds. Lunch would be simple but memorable: dal, rice, pumpkin sabzi and curd—all served with a story.

A few weeks later, the banyan acquired a new tenant, a king cobra. The gardener saw it first, coiled near the roots. Then one day, it appeared on our kitchen steps. By the end of the week, it had made guest appearances in the gardener's quarters, the fish pond and, on one memorable evening, in our veranda.

Panic spread faster than reason. Neighbours advised immediate evacuation. Granny refused. 'Snakes don't come without reason,' she said calmly. 'They remind us to behave.'

She called Major Rajiv Bhatia from the nearby military workshop and requested his 'engineering expertise' to relocate our new resident. The major arrived with two soldiers, a gunny bag and infinite patience. By evening, the mission was complete.

'The guest has been dropped near the University forest,' Granny declared, satisfied. 'It will have better company there.'

From that day onward, all creatures were under her protection. Snakes were honoured guests, fish were auspicious, dogs were wise and humans were, well, tolerated.

Our evenings had a rhythm of their own. When my parents went to their army parties, Granny and I stayed home. We'd switch on our prized possession—a black-and-white television—and watch whatever Doordarshan offered: *Krishi Darshan*, *Chitrahaar*, classical music, the 8 p.m. Hindi news, followed by the same in English. She would knit during news bulletins, critique government policies during the commercials and pray during the national anthem. Once the TV was switched off, she'd settle into her real performance: storytelling.

Her stories began softly, like lullabies, and ended dramatically, often involving ghosts, kings or moral crises. Ghosts were her favourite characters—some benevolent, some mischievous, all full of purpose. 'The ones in trees,' she said, 'are good ghosts. They guard the land. The ones who wander on roads have lost their home.'

It was under her moonlit stories that I learned my first lessons in ecology, long before I knew the word.

If you grew up in India, chances are someone warned you about the peepal tree. It wasn't just a tree; it was the prime real estate for ghosts. Granny confirmed it.

'Never loiter near a peepal after dark,' she'd whisper, 'Baburam and his friends might follow you home.'

Baburam was our household ghost, famous, feared and oddly respected. He was said to live in the highest fork of the peepal, conducting midnight conferences with other spirits. I was fascinated. Why the peepal? Why not the banyan, neem or mango? Perhaps it was because the peepal breathed even at night, releasing oxygen like a saint. Or maybe its leaves, heart-shaped and trembling even in still air, gave an illusion of constant life. Or maybe, as Granny once said with a twinkle in her eye, 'Ghosts like good company, and peepal is the wisest of trees.'

Whatever the reason, the arrangement worked perfectly. The ghosts protected the peepal, and the peepal protected the ghosts. And we humans, we learned not to interfere. It was, in retrospect, a brilliant environmental strategy. 'Long before forest laws,' I often say now, 'we had ghost laws.'

One summer evening, walking with Granny, we passed a cremation ground. A magnificent peepal stood there, its canopy glittering like green glass in the sun. 'Why do peepals grow so well here?' I asked.

'Because here, no one disturbs them. Here, the trees keep company with souls on their final journey. Souls speak the truth,' she said. 'They cry, they remember, they let go. The tree listens.'

To her, the cremation ground was not a place of fear, but of honesty. 'The *shamshaan*,' she said, 'is the only place where everyone bows equally. It's the real temple. No priests. No lies.' She had the clarity to make it simple for a child: '*Mandir toh shamshaan hota hai. Shor-sharaba toh basti mein hota hai.*' Temples are quiet like cremation grounds, she said, while all the noise belongs to human settlements. And indeed, the peepal at the edge of those grounds—majestic, rooted, unhurried—were the real guardians of silence. Years later, when I planted peepal along highways and cremation grounds, I saw how true her words were. Those trees thrived, watered by silence, guarded by memory.

Looking back, I realize Granny's ghosts were never about fear. They were her way of teaching reverence for what we do not understand. To her, the peepal tree was both temple and threshold, a place where the seen and unseen sat together.

'Fear teaches you manners,' she'd say. 'You don't talk loudly under a peepal, you don't lie near the river, you don't waste food in a kitchen. Ghosts are just reminders to behave.' Her wisdom worked better than any moral science class. And her humour . . . that was her secret fertilizer.

By the time we left Range Hills Road, two thousand trees had found a home in the cantonment. They stood quietly by the roadsides, in empty lots and behind bungalows. When we drove away, I looked out of the jeep and felt a strange mix of sadness and pride. The saplings waved in the wind like children left at a boarding school, anxious but hopeful. Granny squeezed my hand and said, 'Don't worry, Munna. The ghosts will take care of them.' I believed her then. I still do.

Even now, when I pass by an old peepal, its leaves shimmer like small green tongues of memory. I pause, bow slightly and listen. Somewhere in that rustle, I hear Granny's voice, equal parts affection, mischief and command.

'Keep planting, Munna,' she seems to say. 'The ghosts are counting on you.' And I smile. Because after all these years, I've realized something. Ghosts never really leave. They just grow roots.

Chapter 4

More Lessons from Granny

Granny filled my world with stories. She spoke of Gods and ghosts as if they were neighbours, never distant, never abstract.

To me, the ghosts were never frightening. They were not eerie visitors of the night. They were companions and comforting presences. I grew up seeing them as guardians perched on branches, invisible yet present. Whenever I felt lost, I would run under a peepal, beneath its canopy, into the circle of quiet, and talk to them aloud without fear of interruption. They were my silent council. They never judged, never answered back, but always lifted the heaviness in my heart. That space was more sacred than any hall of worship, because it was entirely my own. I came to believe that ghosts do not live on trees; they live in us, in our thoughts, our doubts, our guiding voices. We call them ghosts because we cannot name them otherwise.

Granny gave them form and meaning. She insisted they were not to be feared but befriended. 'They are our ancestors,' she would whisper, 'watching, guiding, protecting.'

The village elders had their stories: Gods in flowering trees, Gods in fruiting trees, spirits choosing their canopies. Granny cut through the confusion: 'Stick with the peepal,' she said. It was the most loved, the most enduring. I trusted her instincts. Over time, I saw that her wisdom was practical. Temples and sacred spaces protected peepals. They survived where belief sheltered them. Faith, even in its superstition, had a conservation instinct hidden within. I never laughed at those beliefs. Even if the reasoning felt magical, the result was real: the trees lived. And around them, so did the rituals of *parikrama*, red threads, water offerings, lamps and whispered prayers. People tied their hopes to the branches and, in return, the trees carried their sorrows quietly.

Granny often said, '*Brahman hi Brahm door karta hai*' (Travel removes ignorance).' Her wisdom meant that the divine alone removes the distance between you and truth, and travel helps that understanding. She told me to travel all my life so I might understand the world better.

Those early years of belief, half myth, half memory, were not wasted storytelling hours. They were my apprenticeship in devotion. Granny's rituals were my first lessons in fieldwork. She never called it that, of course. To her, it was simply doing what must be done with love. When she bent to light a lamp or sprinkle water at the doorstep, she was teaching me concentration, humility and the rhythm of care. I didn't know it then, but she was training me to plant trees. When I began digging small pits along Range Hills Road, I found myself repeating her gestures. Each sapling I touched felt like an extension of her world, a living echo of her prayers. The earth was her altar; I was her apprentice priest. Every pit became a prayer, every drop of water an offering.

My red cycle stood nearby like a temple bell on two wheels. The sound of its chain reminded me of her voice calling me for supper. The mud on my fingers replaced the sandalwood paste she once dabbed on my forehead. The rituals had merely changed form; the devotion had stayed the same.

Some days, I pedalled through heat that shimmered like molten brass. The tar roads smelled of exhaustion; the saplings trembled on the back carrier like nervous passengers. I called them by name, 'Come on, Peepal, don't faint before the water arrives!' Balancing bottles of water was an act of acrobatics. If I leaned too far left, the bottles spilled; too far right, the cycle groaned in protest. Once, the cork came loose, and the whole supply vanished before

I reached the site. I wanted to cry, but Granny's voice drifted through memory: *'Patience is the first prayer.'* So I picked up the empty bottle and promised the saplings I'd return, and I did.

Each evening, when the breeze softened, I checked on them. Some drooped, others stood defiant. Every survivor felt like a victory. I began to understand what Granny meant when she said, *'You don't force life to grow, you accompany it.'*

People began to notice. Children followed me like squirrels chasing nuts. Old women smiled indulgently. A few cynics muttered, 'Goats will eat them by morning.' One laughed outright. 'You're wasting time!' he said. I smiled. 'Then I'll plant for the goats too,' I replied.

Shopkeepers began to recognize me. One Parsi uncle from the cycle shop winked and said, 'Arrey, gardener boy, you're greening the road faster than the government!' Another handed me a handful of mud as *prasad*. That's when I understood what Granny had done: she had blurred the lines between Gods and ghosts, between the living and the unseen. My field was full of both, saplings whispering like souls waiting to be remembered.

When the rains came, I rode through puddles as if the monsoon itself had become my accomplice. The cycle squeaked in protest but carried me faithfully. Sometimes I spoke to it aloud: 'Hold steady, partner, we have a forest to grow.' People stared, but I had long stopped caring. Faith never looks entirely sane. For Granny, it was Krishna stealing butter. For me, it was a red cycle carrying ghosts that would turn into trees.

Each new leaf that unfurled felt like her smile returning through the branches. The more I planted, the more I understood her invisible world. She saw spirits in the peepal tree; I saw spirits in every green shoot breaking through the earth. We were speaking the same language; hers in chant, mine in chlorophyll.

By the time my father's transfer to Jalandhar came through, Range Hills Road had changed. The once-dusty stretch now shimmered with tiny saplings standing like schoolchildren at morning prayer. That sight gave me a satisfaction no report card ever could. On my last evening before leaving Pune, I cycled one final time down the road. The air was thick with the smell of wet dust. I stopped by the first peepal I had planted. It reached my shoulder now, its leaves trembling like small green flags. 'You'll be fine,' I whispered. 'Don't miss me.' The wind rustled in reply, and I could almost hear Granny's chuckle: *'You planted faith, not trees.'*

When we reached Jalandhar, I carried that faith with me like a secret seed. Granny had already taken root within my spirit, but in that new soil, it needed watering. And as always, life sent a teacher to show me how to grow it.

Looking back, what people called innocence was never weakness; it was strength. I had no grand plans or offices, only conviction and simplicity. That disarmed people. They could not refuse a boy asking not for himself but for a tree. Innocence is contagious. A coin given did not vanish; it became a root. A bucket of water became shade. Once people saw that, they kept giving. That was my discovery of micro-donations—each small act a thread in a vast web of trust.

Returning years later, I stood beneath peepals I had planted as a boy. Children who once helped me water them had become parents, resting in their shade. The work had become larger than me, sustained by patience, innocence and faith.

Our awards and rewards do not come from humans or institutions. They come from birds and insects, the real auditors of our work. Their return is our certificate of success.

Granny's presence was more than stories; it was discipline disguised as affection. She taught me to make nursery beds, to value simplicity, to trust people, especially women, for their strength and generosity. She showed me that cleanliness was not only for the house but for the mind: *'Keep your cycle clean,'* she told me, *'it is your life partner. It carries you, it serves you. Respect it.'* That lesson stayed long after childhood.

Her life was austere, yet regal. She lived with minimal possessions, but her heart was expansive. She showed me that generosity is not about wealth but about presence. Even in her frugality, she made room for others. That spirit shaped me more than I realized at the time.

Living with Granny was like inhabiting two worlds at once: the visible world of chores, school, meals and visitors, and the invisible world of Gods, ghosts and whispered guidance. The two never contradicted each other; they completed each other. She showed me that ghosts could be angels, that superstition could be protection and that faith could be as practical as a hand on your shoulder.

When I think back, I realize Granny was the bridge. She taught me how to live among people without losing touch with the unseen. She kept one hand in the world of Gods, another in the world of soil, and placed me gently in between.

While others sought masters in distant caves, I found mine under the leaves of a ficus. Granny's lessons allowed me to realize that the *guru within* awakens when we allow space for silence. Gandhi's spinning wheel taught me that small acts, done sincerely, weave a larger fabric. Rajneesh showed me that courage lies in asking questions. Between them, I learned balance, between action and contemplation. The ghosts Granny spoke of, became, in my mind, gentle guides whispering patience and faith.

Every sapling I planted was also a seed within. A wilted plant taught me neglect; a thriving one taught me resilience. Seasons became sermons. Rainfall became scripture. Over time, I understood: the truest teacher is inside us. Books, saints and stories only light the path; the walking must be done within. The *guru within* is the same intelligence that moves rivers, grows forests and carries life forward.

My field notes became my compass. I wrote in diaries during rides and roadside halts, thoughts arriving under trees, not inside walls. Consistency mattered more than enthusiasm. Patience mattered more than slogans. My accountability was sacred; every tree was personal.

Returning years later to the highways, I found those trees standing strong, some now worshipped, some shading travellers. The work had outgrown me. The trees no longer carried my name; they carried the community's memory.

Chapter 5

Mrs Williams's Classroom

It was around this time that those questions began to find their first answers. If I close my eyes, I can still hear Mrs Williams' voice rising above the rustle of the leaves by the Ram River. It wasn't just a teacher's voice. It was the kind that could make you stop mid-scribble and listen, as if the wind itself had something to say. That voice taught me to see the invisible, to listen to the hum of the earth, to the gossip of leaves, to the quiet diplomacy between roots and stones. It was in her classroom at Spicer Elementary School, Pune, that I first learned to read the language of life beyond the textbook.

I had entered Spicer in the fourth grade. My father managed my admission under the defence concession, and in a matter of weeks, I found myself promoted to the fifth grade. For months, I strutted around with the swagger of a genius. I believed that the teachers had recognized my brilliance, my uncommon grasp of the alphabet and my superior understanding of multiplication tables. It was only later that I discovered the truth: it was simply the privilege of being an army officer's child.

At that age, even illusions made one feel special. I wore mine like a medal.

The years 1974 to 1978 were my Spicer years—formative, free and fragrant with discovery. The school was part of what is now Spicer University, then a modest outpost of Andrews University, Washington. It sat on the green edge of Pune's Aundh village, where bullock carts still ruled the roads and the afternoons smelled of sugarcane and drying hay. The gentle Ram River wound through the campus like a line from a poem—slow, silvery, self-assured. To the east lay the University of Poona, whose domes gleamed in the sunlight; to the west, golden fields stretched lazily into the horizon. Between these two worlds, one of scholarship and one of soil, stood our small school, sturdy and unpretentious, a sanctuary for curiosity.

She came from Sri Lanka, from a forest-dwelling tribal community, we were told. She was tall and angular, with frizzy hair cropped close like a halo of discipline, deep-set eyes that could read both your notebook and your thoughts, and long, tapering fingers that seemed designed to point out the secrets of the world. Her skin bore the bronze gloss of sun and labour; her walk was that of someone who had never learned to be tired. To us, she was a wonder: part teacher, part magician, part storm.

Mrs Williams never carried a textbook. She appeared to know everything: geography, history, science, grammar, weather, even the unspoken politics of playground friendships. If she lived in this age, we would have called her *Mrs Google*. But she needed no search bar; her memory was a rainforest—vast, teeming, endlessly alive.

She had a soft laugh, like the crackle of dry leaves after rain, and a sense of humour that could make even the strictest principal smile. She never raised her voice. Her authority came from our curiosity. We obeyed her because we didn't want to miss what came next. When she spoke of Argentina, you could almost hear the tango; when she mentioned Antarctica, the classroom grew colder by a degree. Her words travelled faster than aeroplanes, and she landed in our minds long before we had passports. Her life in the jungle explained her oddities, the ease with which she handled snakes, her habit of predicting rain by scent, her insistence on walking barefoot for 'better learning conductivity'. She said knowledge travelled faster through the soles. She made us remove our shoes and stand barefoot on the earth. 'Learn to feel the ground,' she would say. 'Knowledge begins with touch.'

One morning, she asked us to bring clean jam bottles from home. Inside them, she said, we would carry our lessons. By the following week, our classroom looked like a miniature zoo. Children arrived with jars holding beetles, grasshoppers, snails and occasionally a startled lizard.

She showed us how to hold them gently, observe, and release them unharmed. My parents were horrified. My grandmother, however, clapped her hands in approval. 'At last,' she said, 'you've found your tribe!'

Mrs Williams took us out of the classroom often, to the rocky bends of the Ram River, to the fields where dragonflies shimmered like coloured glass, to the tall grass that whispered secrets. That riverbank became our second classroom. We learned how tadpoles vanished into frogs, how snails sealed themselves in summer, how catfish hid in the mud waiting for the monsoon. The school bell was replaced by birdsong, and our textbooks fluttered in the breeze like uncertain butterflies. Parents, of course, were scandalized. A few marched to the principal to complain. The principal, also a Sri Lankan, simply smiled. 'Let the earth teach them,' she said. The complaints dissolved like chalk in rain.

We learned to carry our secrets home quietly. Our pockets held pebbles, seedpods and sometimes wriggling things that later escaped during dinner. Once, a frog jumped from my tiffin box right onto the dining table. Granny took it as a divine sign. My mother nearly fainted. My father threatened to send me to a boarding school in Rajasthan, 'where nothing green survives'.

Under Mrs Williams, we became small farmers. Our homework was to plant and nurture something—tomatoes, spinach, beans—and report progress. By the end of three years, every child had a balcony garden or a small patch behind the house. Some of us may have failed in algebra, but none failed in germination.

My father, always serious about education, once took a loan of Rs 13,000 to buy a sixteen-volume set of *The Standard Britannica Encyclopaedia*. Mrs Williams had insisted I read beyond textbooks, and my father, convinced I was destined to be a scientist, obliged. Those heavy volumes sat on the shelf like patient monks waiting to be read. I, however, preferred the encyclopaedia of soil, where every earthworm was a paragraph and every pebble a punctuation mark.

Before Mrs Williams, life had been peaceful and predictable. After her, it was crawling, buzzing and gloriously untidy. She had turned me into a keeper of biodiversity. My schoolbag carried notebooks and jam jars; my room smelled like the forest floor.

We learned big words early: *interdependence*, *ecosystem*, *decomposition*. We couldn't always spell them right, but we could demonstrate them with insects. 'Everything depends on everything else,' she would say. 'That's how the world works.' Those words became the roots of everything I would later believe about ecology, society and even destiny.

Mrs Williams was obsessed with handwriting. She demanded perfect cursive. 'If your letters don't flow,' she said, 'your thoughts won't either.' At first, our pages looked like wounded soldiers. But within six months, our notebooks were works of art. That discipline became the foundation of my diaries—those endless pages of ink where I now plant memories instead of saplings.

Sometimes she would slip in philosophical grenades mid-lesson: 'Natural selection extends to the choices we make,' she'd say, adjusting her glasses. 'Creation has no purpose; it simply expresses.' We nodded solemnly, pretending to understand. Decades later, those lines would germinate in me like seeds buried deep, surfacing at the oddest times.

Among our most cherished possessions was a magnifying glass. Mine had a red rim and cost ten rupees. Through it, I discovered the geometry of miracles: the veins of a leaf forming perfect architecture, the crystalline structure of sand, the spirals of a snail shell. Under its circle of light, the world revealed divine symmetry. If there was perfection anywhere, it lived in these small, unadvertied corners of creation.

One morning, after a story on conservation, Mrs Williams looked around and said, 'Children, trees are not decoration. They are the breath of your home. Each of you will plant one this week.' Something inside me leapt. I had been secretly doing this already, stealing time, collecting saplings, digging small holes near drains and temple courtyards. Now, suddenly, it had a name. Mrs Williams smiled, as though she had been waiting for this confession. 'We'll call our group *Give Me Trees*,' she said. And just like that, a seed sprouted into a movement.

Our classroom transformed into a nursery. Pots, polythene bags, discarded tins, everything was used.

The smell of wet earth replaced chalk dust. The corridor bloomed with hibiscus, marigold, lemon and guava. Gardening became a lesson in science, patience and poetry all at once. 'When you touch soil,' she said, 'you touch history. When you water a plant, you are writing tomorrow. Her words echoed Granny's, though in a different accent. Granny spoke of spirits in trees; Mrs Williams spoke of systems and cycles. One had given me roots, the other gave me direction. One taught me faith, the other gave me language.

That was how *Give Me Trees* was born, not as an NGO or an organization, but as a prayer in uniform. A grandmother's whisper had turned into a classroom pledge.

Even now, decades later, I sometimes pause mid-planting and hear her voice beside the river: 'Learn to feel the Earth. Knowledge begins with touch.' I realize she never just taught us subjects. She taught us reverence. She gave us curiosity, turned soil into scripture and every jam bottle into a temple of observation. She never claimed credit. She never lectured about success. She simply lived what she taught—quietly, fiercely and joyfully. If Granny sowed the seed, Mrs Williams watered it. One wrote my prologue; the other began my story. So, when people ask where *Give Me Trees* began, I smile and think back to that bright morning by the Ram River, the barefoot teacher with frizzy hair and a voice that could stop a breeze, holding up a leaf to the light and saying, 'Look closely, children. The Earth has handwriting.'

That was Mrs Williams's classroom, where nature first began to speak, and I first learned how to listen.

Chapter 6

The Roadside Gardener

It was the last week of January 1977. Twenty-sixth January, my eleventh birthday. My Granny had planned my baptism long before I knew what the word meant. Not the kind performed with holy water, but with soil and saplings. When I returned from the Republic Day parade at school, still clutching the tricolour paper flag and humming the national anthem, she was waiting at the gate. Birthdays in those days were simple affairs. There were no cakes, no balloons, no deafening music. Just a plate of halwa-poorie, a few friends and Granny's soft humming from the kitchen. That day, there was also cold coffee and potato wafers she had made at home—our luxury.

Sunil Kaka, our gardener, took me on his old bicycle to a small nursery on the edge of Kirkee Cantonment. The nursery smelled of damp soil and cow dung, and the owner greeted us warmly. After some cheerful chatter in Marathi that I half understood, he handed over nine saplings: two peepal, two neem, two banyan, two jamun and one goolar. I held them as if they were living secrets. We rode back to our bungalow, T-23 on Range Hills Road, where Granny waited like a general inspecting her troops. She had already marked the planting spots, six bungalows long, around our neighbourhood. That evening, we walked along the roadside, the light fading into gold, the smell of rain-soaked dust still hanging in the air. She told me where to dig, how deep to plant and how firm to press the soil. I planted my first tree—a banyan—in front of bungalow number T-20.

Forty-eight years have passed. Those nine saplings stand tall today, majestic canopies sheltering birds, squirrels and memories. The banyan and peepal I planted have grown into landmarks of their own, their roots gripping the road like veins of time. Whenever I visit, I stop to touch their trunks, the same hands that once pressed their roots into soil now tracing the rough bark of adulthood. That was my baptism. That was the day I became a roadside gardener. The roadside became another classroom. There were no benches or chalkboards, only the hum of buses, the curse of drivers and the quiet patience of the earth. My lessons began in 1977—the year a red bicycle entered my life. It was a Speedking, shining like laughter under the Pune sun. My father bought it from the British Cycle Company opposite Kohinoor Café, a shop that smelled of grease, oil and Parsi generosity. I was eleven, and that bicycle gave me wings.

Granny stitched thick cotton bags from old curtains for me and cleaned old oil cans for water so I could carry ten litres each trip. I tied the cloth bags to its handlebars, filled them with saplings rescued from walls, drains and temple courtyards, and set off down Range Hills Road, carrying baby trees like companions, in search of lonely soil. At that age, few parents would have allowed a child to cycle alone on highways, but my parents and grandparents did. Those were different times—safe, slow, kind. Roads were ruled by rickshaws and bicycles, not speed. I didn't know it then, but I was already chasing my life's calling, not as a plan, but as an instinct.

Funds, of course, were scarce. My pocket money disappeared like morning dew. Granny contributed wisdom and an occasional ten-rupee note. My mother gave her old sarees; we tore them into strips to tie the saplings. When even that ran out, I discovered the oldest currency on earth: goodwill. Strangers offered spades, gardeners lent manure, shopkeepers gave water in discarded bottles. Someone once handed me a slice of mango and said, 'For your tree also.' I realized generosity grows faster than grass when watered with purpose. I planted mostly near temples, akharas and dharamshalas—places where reverence ensured survival. At first, priests objected: 'Peepal grows too big, beta!' But when they saw my persistence, they relented. 'Fine,' one said, 'I'll water it.' That was all I needed.

Years later, those saplings stood tall, each with its small human ecosystem beneath. Tea stalls, rickshaw drivers, beggars, lovers, preachers, everyone found shade there. I understood then that a tree never grows alone. It pulls an entire civilization towards itself.

Loss came too. Goats chewed, children uprooted, storms swept away. Each death hurt. But Granny consoled me with her timeless remedy: 'If one dies, plant two. Let the Gods do the counting.' She was right. I stopped mourning, started multiplying. My red Speedking became my passport to freedom. Mud on my knees, leaves in my schoolbag and the smell of wet earth on my hands—that was my childhood uniform. My father received complaints: 'Your son is digging up the cantonment again!' But he would only sigh and smile. 'Some cases,' he'd say, 'have no cure.'

Fieldwork was never glamorous. It was sweaty, muddy, mosquito-friendly, heartbreak-prone. Sometimes, the only applause came from crickets. But it was full of laughter. A goat once ran away with my measuring tape. Another time, a policeman asked, 'Do you have permission for this?' I said, 'Sir, I'm helping the government increase green cover.' He stared, then muttered, 'Next time, carry your nursery license.' I nodded, pretending such a thing existed. Gradually, my mission as a roadside gardener sprouted legs of its own. Housewives began carrying saplings in handbags; shopkeepers kept water pots; schoolchildren volunteered as tree guardians. Two people on a cycle or a scooter with ten saplings became a familiar sight. We called ourselves a *groundforce*, though most of the time we were just two people and an enthusiastic dog.

My diaries became my compass. Date, species, weather, remarks. Pressed leaves, soil samples and even notes on the ants that visited. 'Write as if the soil is reading,' Granny used to say. Those pages became my degrees, earned in mud, not marble.

At dawn, as I sip tea, I hear Granny in the breeze: '*Keep your windows open; ideas need air*.' And I realize: windows too are classrooms. They teach perspective and patience. I always wanted the window seat, in trains, buses and life itself. The world outside was always my teacher.

People often tell me, 'You must be proud, millions of trees, thousands of volunteers.' But pride is brittle. I prefer gratitude; it has better roots. My real joy lies in knowing that somewhere a stranger rests under a tree I once planted. He doesn't know my name; I don't need to know his. The shade is shared. Democracy achieved, silently. Granny taught me that gardening is not a profession; it's a philosophy. To plant is to believe in tomorrow. To water is to confess faith in continuity. To compost is to make peace with decay.

And so, I continue—older, slower, but still a roadside gardener at heart. I no longer count saplings; I count mornings. Each dawn is a new planting. I still believe that the earth listens, that dust remembers, that even ghosts need shade. When I die, I hope someone finds a neglected patch of roadside and plants a sapling, quietly, anonymously. No plaque, no name. The tree will know. It will grow, give shade, host birds and ghosts alike. And in that quiet generosity, I shall live on.

Chapter 7

Seeds of an Idea—*Give Me Trees*

Things were looking good, and it was time to harvest. I was developing confidence with saplings, my Granny by my side, my parents ready to support me, my school cheering me on, the neighbourhood rising in chorus and even the nursery owners amused by my enthusiasm. All I needed was a mission that would give me direction and momentum.

Mrs Williams gave me that before we left for our new posting. My father, having completed his MD in Paediatrics from the Armed Forces Medical College in Pune, received a promotion and was posted to Jalandhar. Before I left, Mrs Williams gave us something precious when she formed an eco-club for our class and named it *Give Me Trees*.

The idea was simple: create saplings, plant them, care for them, green our homes and even grow vegetables as homework. I took to it instantly. I was far ahead of my classmates, and Mrs Williams knew it.

When my father moved to Jalandhar Military Hospital, we followed: me, Granny, Mom and all the luggage of our small household. I carried the idea of *Give Me Trees* with me like a sacred seed. Mrs Williams moved elsewhere, my friends drifted into their studies, but *Give Me Trees* (GMT) was not dead; it lived in me. I began planting under that name. It was easy to explain to people what I was doing and why I needed their support. The name said it all: *Give Me Trees*.

And *Give Me Trees* never looked back. As I grew in height, so did my plants. As I entered my teens, GMT blossomed, and so did the trees. Even today, the blessings of that idea are with us, and I am sure Mrs Williams must be smiling down from the heavens.

Soon, the plantation of peepal and neem saplings came under GMT's umbrella. We painted posters, made little flyers and printed simple call-to-action leaflets, asking people to join our mission, to create shade and habitats for birds. *Give Me Trees* has taught me endlessly. It showed me that every journey has an outer path and an inner path. The roadside, the rooftops and the saplings formed my outer journey; but slowly, through people, books, and silences, I began walking the inner one.

Predators, I learned, are protectors of the forest. Without them, ecosystems collapse. What seems cruel to human logic is sacred balance in nature's code. Understanding this changed how I worked. I also learned that forests heal themselves if given time. Three years of fencing off a patch from grazing can bring back astonishing life. Patience is the real seed of regeneration. Buildings may last centuries, but forests are forever. They were here millions of years before us and will remain long after we are gone.

Give Me Trees is not just an organization; it is a philosophy born of faith, movement and simplicity.

It taught me that a sapling, like a thought, grows best when nurtured daily. Every act of planting is a prayer. Every shade is a blessing returned.

Chapter 8

Soil Matters

Granny used to say, '*Zameen roti hai, zameen roti banaati hai.*' The earth weeps, and the earth feeds. As a child, I thought it was a riddle. Only later, after working with degraded, exhausted soils, did I understand. Everything begins with the soil and ends with the soil. In fact, soil is the only thing to understand. Farming, agriculture, afforestation, social forestry, nature conservation, grasslands, wetlands—whatever you wish to study, you have to first study the soil. That is why they say, 'Dust unto dust.' Soil is the soul of everything. It holds everything—tears, ashes, laughter and loss. It holds the planet's memory, its chemistry, its poetry, its destiny. Every civilization that respected its soil prospered; every one that ignored it perished. The Indus Valley, Mesopotamia, the Mayans—all rose on fertile floodplains and fell when their soil turned to dust. Even today, our future depends not on satellites but on soil samples. Yet, soil has always been the most misunderstood part of creation. People walk on it, build over it, bury their dead in it, but seldom do they stop to touch it, to listen.

When I was young, I planted saplings like a soldier on a mission. I counted numbers, not heartbeats. Survival rates mattered more than soil health. I wanted to see green, fast. But the soil, like an old teacher, began to whisper patience. You cannot rush the earth. If you wish her to bloom, you must first feed her. So, I learnt the science and art of working with the soil, my granny becoming my first soil coach. I began to feed the soil the way a mother feeds a child: patiently, lovingly. I learned to mulch, spreading dry leaves around the base of young trees to conserve moisture and keep the roots cool. I mixed jaggery with cow dung to make *jeevamrit*, that living tonic of microbes. I sprinkled wood ash to balance acidity, poured compost tea to restore bacterial balance and even turned kitchen waste into treasure. Slowly, the soil changed. Cracks softened. Earthworms returned. Rainwater soaked instead of running away. I learnt to heal the soil, and in doing so, I understood something vital—decay is not death; it is rebirth in disguise.

When a tree sheds its leaves, it isn't losing them; it is feeding its own roots. When an animal dies, its body becomes a feast for millions of microbes that keep the wheel of life turning. Death is just the soil's way of saying, *'Don't worry, I'll take care of it.'*

That is why some of the most majestic peepal trees grow on cremation grounds. Their roots feed on ashes; their leaves shimmer with new life. Soil does not discriminate; it absorbs sorrow and turns it into strength. The chemistry of decay is, in truth, the science of compassion.

Soil science may appear complicated, but at heart, it is pure poetry. The chemistry of the soil is the chemistry of life itself. Nitrogen, phosphorus and potassium—the sacred trinity of every farmer's chant—are not just nutrients; they are musical notes in the song of creation. Nitrogen gives the green to leaves, phosphorus fuels the roots and flowers and potassium strengthens the plant's immune rhythm. Around them orbit the minor musicians—calcium, magnesium, sulfur, zinc—quietly keeping time like percussionists no one notices but everyone needs. And then there is carbon, the silent archivist. Every fallen leaf, every dead root, every worm that wriggles and dies becomes a line in this vast ledger of memory. Over time, it turns into humus, that dark, fragrant miracle that keeps the soil alive. Science calls it organic matter. I call it forgiveness.

I learned more chemistry through blisters than through books. When I mixed cow dung with dry leaves and kitchen waste and watched the pile heat up, steam and cool again, I realized I was watching alchemy in action. Composting is the poetry of decay, the transformation of yesterday's waste into tomorrow's fertility. The day I smelled my first batch of homemade compost, I knew what creation must have smelled like when it began.

I became addicted to the smell of restored earth. To kneel beside a patch of land once barren and see it darken with humus is better than any medal or degree. Once nourished, irrigated and loved, the soil does the rest on its own. Seeds know what to do. Saplings find their rhythm. The soil takes over its own healing.

Dig deep enough, and you find not just roots, but relationships. Beneath every tree lies a civilization—fungi, bacteria, nematodes, protozoa—all living, trading and conversing. Fungi form filaments called *mycorrhiza*, tiny lifelines linking one root to another, carrying nutrients like postmen of the underworld. Trees send messages through them, warning neighbours of drought, sharing sugars with the weak. Science calls this the underground internet of roots. Granny would have called it *prana*, the flow of life. A single teaspoon of healthy soil holds a billion living beings. Invisible, indispensable. When we pour chemicals, we massacre them. When we mulch and compost, we invite them home. They are the quiet engineers of fertility, the real guardians of our forests.

Once I understood their world, I began to plant differently. I realized I wasn't just growing trees, I was healing the soil and, in doing so, dealing with something within myself. I stopped disturbing the soil too much. I let leaf litter stay. I planted in clusters so roots could hold hands underground. Every successful plantation I have done is the success of that invisible community, not of man's labour alone. Understanding this at an early age allowed *Give Me Trees* to be born and to survive.

I have worked on many wounded lands—burnt, compacted, scraped, abandoned. Each had its own illness. They looked dead, like deserts of dust. But soil never truly dies; it merely sleeps. You just have to wake it gently. I would ride for hours with gunny bags of compost and a few trembling saplings tied behind. I learned to read soils by sight and smell. If the earth looked dull, it needed carbon. If it smelled sour, it needed air. If it cracked, it needed mulch. The soil spoke in textures and tones, and my job became to listen. Every patch of ground had its temperament: the sandy soils that let go too easily; the stubborn black cotton soil that clung like memory; the red laterite that bled rust and iron. I got to know them all. I've worked with them, argued with them and finally learned to love their moods. Soil was both patient and physician. It taught me that we cannot command the Earth, we can only collaborate with her; that every act of planting is also an act of surrender. Once the soil healed, it healed everything else—water, air, plants and even people.

If we heal the soil, we heal the climate. If we feed the Earth, we feed ourselves. The planet does not need more developers; it needs more decomposers. In the end, soil is not just material. It is a metaphor, a mirror and a mother.

It forgives quietly and waits for us to return. When we kneel beside the soil, something ancient awakens. The scent of damp earth after rain, that first curl of petrichor, is not just fragrance. It is the voice of microbes rejoicing after a long sleep, the earth exhaling after holding her breath too long. When I dig a pit now, I no longer see it as work, but as prayer. My hands go into the earth, and the earth, in return, enters me. To understand soil is to understand everything else: the trees, the rivers, the seasons and the human spirit. And one day, when I too become part of this soil, I hope someone will plant a sapling there, just to keep the conversation going.

Chapter 9

The Miracle of Composting

For years, I have watched the Earth digest everything—leaves, memories, mistakes, even grief—and return it all as nourishment. I have begun to think of life itself as one long act of composting. From the day we are born, decomposition begins, not as decline, but as transformation. The body, like the Earth, is a biochemical experiment in renewal.

When I dig into a compost pit, I am not just turning waste; I am turning time. The moment organic matter begins to decay, millions of unseen workers awaken: bacteria, actinomycetes, fungi, protozoa, nematodes and earthworms. Each has a role in this theatre of transformation. The carbon-rich materials, dry leaves, sawdust and paper meet nitrogen-rich companions like vegetable peels, grass and manure. Carbon provides energy; nitrogen provides protein. Add oxygen and moisture, and the orchestra begins!

Temperatures rise as microbes feast, sometimes reaching 60°C, sterilizing pathogens while weaving new life into humus. That dark, crumbly material called compost is no small miracle. Chemically, it is a stable colloidal complex, a molecular mosaic of carbon chains, organic acids and minerals. It binds soil particles together, enhances porosity and nutrient absorption capacity, and holds moisture like memory. A teaspoon of healthy soil can host more microorganisms than there are humans on Earth. To kneel beside it is to touch an entire civilization invisible to the eye.

When I first learned to compost, I thought it was an act of cleaning. Later, I realized that it was an act of prayer. Composting is the science of humility; literally, *humus* is the root word of *humility*. Every peel, leaf or dead flower carries its potential in carbon bonds waiting to be rearranged. The process is pure chemistry:

- $C + O_2 \rightarrow CO_2$ + energy (the breath of decomposition)
- Proteins → amino acids → ammonia → nitrates (the nitrogen cycle in miniature)
- Lignin + cellulose → simple sugars → humic acids (the long patience of soil alchemy)

These are not mere reactions; they are revelations. The Earth does in her pits what we must learn to do in our hearts: break down rigidity, release energy and become fertile again.

Over the years, I have observed the temperament of soil the way one studies friends. The red laterite soils of Pune are moody, rich in iron oxides, quick to drain and ideal for neem and jamun. The alluvial soils along the Gangetic plains are generous, fine-textured, high in silt and friendly to roots. The black cotton soils of Jalandhar cling stubbornly, their clay particles swelling with the monsoon and cracking open in summer. Each has its personality and memory. Soil, like humans, responds to how you treat it. It can become resentful, compacted or lifeless if suffocated under cement or chemicals. But the moment you loosen it, feed it compost and let it breathe, it revives.

Granny, without knowing a word of microbiology, understood all this intuitively. '*Mitti jeeti hai*,' she used to say, *the soil is alive*. When I dug pits with her as a child, she would add cow dung and leaves in layers, muttering small folk mantras of gratitude. She was performing, unknowingly, a lesson in the carbon-nitrogen ratio, roughly 30:1, the golden balance of decomposition. Too much nitrogen, the pile stinks; too much carbon, it sulks. In life, too, I've found that balance is everything: enthusiasm must meet patience, action must meet rest. Compost is, in essence, condensed wisdom. Its benefits are not poetic imagination but measurable science.

- It improves soil structure, creating aggregates that allow air and water to flow freely.
- It buffers pH, making acidic soils sweeter and alkaline soils gentler.

- It releases nutrients slowly, ensuring plants sip instead of binge.
- It increases microbial diversity, the unseen democracy beneath our feet.
- It sequesters carbon, turning climate guilt into ground fertility.

In every pit I've turned, I've seen not just waste turning to wealth, but ego turning to empathy. Watching decay taught me more than any classroom did. I began to see how nature manages its economy flawlessly, nothing wasted, everything transformed. Even a fallen leaf is an accountant balancing the books of existence. Each time I crumble compost between my fingers, I feel the pulse of photosynthesis running backwards—a reverse song of life returning to its source.

Scientists speak of Cation Exchange Capacity (CEC) and Humic Substances, of N:P:K ratios and microbial respiration rates. I speak of gratitude, surrender and patience. But in truth, we are saying the same thing. Soil science and spirituality are twins separated by language. Both teach that transformation happens in darkness, invisibly and over time.

When I return compost to the roots of a peepal tree, I feel like I am returning memories to their maker. The soil takes them in silently, mixes them with the remains of other lives and produces something richer than any one life could make alone. That is the essence of the Earth's chemistry: collaboration through decay.

I once kept a handful of compost in a glass jar beside my writing desk. Its fragrance, faint and earthy, reminded me that everything I create will someday crumble, and that this is not failure, but fulfilment. We are all part of one vast metabolic system. Our thoughts, actions and even our stories are nutrients in the making. So, when I look at an old compost heap, steaming slightly on a cold morning, I no longer see waste. I see intelligence at work—ancient, precise and kind. I see the same process running through my veins, in my bones, in the slow ripening of my understanding.

Chapter 10

Rivers of Learning

If the Kumbh was a spectacle, then the small rivers and village ponds were daily soap operas—no less dramatic, just with smaller audiences. Every mohalla had its local water body, and ours was a pond ringed by neem and keekar, buzzing with dragonflies. Buffaloes claimed it as their private spa, standing in the shallows with eyes half-closed, tails flicking like lazy metronomes. Children, of course, clambered onto their slick black backs, shrieking with laughter until the buffaloes sank deeper, leaving us flailing in muddy water while elders shouted warnings we never obeyed. Granny sat on the bank, unimpressed, muttering, 'These buffaloes meditate better than half the sadhus you will meet.'

In the monsoon, small nalas appeared out of nowhere, carving temporary rivers through fields. We built paper boats from our school notebooks, launching half-finished arithmetic problems and teacher scoldings into the current. Races ended in tears when a boat capsized or got stuck in weeds, but also in jubilation when one sailed away triumphantly, carrying dreams we were too young to name. I remember chasing one boat so far downstream that I lost my slippers in the slush and returned home with cuts on my feet. Granny, seeing my sorry state, didn't scold. She only said, 'Beta, the river gave you freedom and took your slippers as tax.'

As I grew older, my world expanded from ponds and nalas to rivers with names that shaped civilizations. The Ganga, Yamuna, Ghaghara, Gandak, Son, Chambal and Betwa—all part of one vast, breathing organism called the Ganga River Basin. Stretching across eleven states, this basin is India's living geography lesson, covering over a quarter of the nation's land and feeding nearly half its people.

Here, rivers are not lines on a map; they are arteries of survival. They irrigate fields, recharge aquifers, move silt, host fish and carry stories older than empires. Science calls it a watershed; our ancestors called it a pilgrimage. Granny called it *Gangaji ki leela*. To her, every tributary was a child of the Ganga, each with its own mood and purpose. She would say, 'The Yamuna is quiet like a saint, the Ghaghara is playful like a child and the Son is temperamental like a king.' Her stories were less geography and more personality studies of water. When I later studied river ecology, I realized she was right. Each river truly has its own temperament, its pH, mineral content, turbidity and flow patterns, creating distinct ecosystems. Science had finally caught up with Granny's folklore.

Walking along riverbanks became my education long before universities took me in. I learned how the colour of water changes with soil type: red in laterite, black in basalt, pale gold in alluvium. I noticed how reeds grow in clusters where the silt is rich, how dragonflies vanish when pesticides enter the stream and how frogs sing before rain because they can sense drops forming in the sky. These lessons came without classrooms or degrees. The rivers were professors; the banks were their blackboards.

Wherever a river flowed, a peepal tree was never far behind. Sometimes a temple stood beneath it, sometimes a single stone daubed with vermilion served as deity enough. These little shrines, half-eaten by time and moss, taught a silent theology of continuity, shade and surrender. Under those peepal trees, I saw old men reciting verses, washerwomen gossiping, saints meditating and children climbing for fun. Ecology, economy and faith, all three met under one canopy. That was India's real open university.

When I walked along the ghats of Allahabad, I often saw how rivers define not just geography but culture. The city itself is a confluence of rivers, religions and realities. The Sangam, where the Ganga and Yamuna meet the mythical Saraswati, is a classroom of coexistence. Every morning, priests perform rituals beside dhobis, and both are watched by students like me, sipping tea, lost in thought. The Ganga there looks weary yet dignified, carrying the burden of faith, industry and indifference all at once.

In Benares, I learned another lesson. There, the river is not worshipped *as* the divine; it *is* the divine. From Assi to Dashashwamedh, every ghat is a chapter in the same scripture. You can read history in the worn steps, theology in the chants and sociology in the garbage. People bathe, pray, sing, cry and die, all within the same stretch of flowing water. The river holds no bias. She carries ashes and offerings with equal grace. Standing on the ghats at sunrise, I felt both small and infinite, as if every breath of mist carried an invisible mantra: *Flow, even when the world doesn't deserve you.*

The science of rivers is, in truth, the science of patience. A river does not resist obstacles; it learns to curve around them. It infiltrates the ground, recharges wells, nourishes crops and feeds unseen aquifers deep below. The Ganga Basin alone supports over four hundred million people, not through miracles but through meticulous natural engineering. Its floodplains store water like giant sponges, releasing it gently through dry months. Its sediments replenish fertility year after year. When we pave over those plains or dam the tributaries without foresight, we break ancient circulations the planet spent millennia designing. And yet, despite our interventions, rivers continue to teach humility. Floods in Uttarakhand and Himachal remind us that water does not forgive arrogance. Rivers will reclaim their paths if denied. Granny used to say, 'Never sleep on a dry riverbed, beta. Rivers remember their way home.' Science later confirmed that floodplains always mark past flows. Her village wisdom had predicted hydrology.

Each river carries its own curriculum: geography, geology, biology and mythology blended into one syllabus of survival. You can learn soil erosion by watching a monsoon bank collapse, learn aerodynamics from the flight of kingfishers and study chemistry from the changing colours of the current at dawn. Every ripple is a textbook page waiting to be read with your feet, not your eyes.

In my years of planting along the Yamuna and Ganga, I have often been asked: Why here, when the soil is saline, the water dirty and the banks neglected? My answer is simple. Planting along rivers is not horticulture; it is repentance. Every sapling is an apology. Peepal and neem, jamun and babool, each one a green prayer for forgiveness. Many were swept away by floods; many survived, gripping silt like faith itself. In Allahabad, I watched the same peepal I planted lean over the water after ten years, its roots embracing the embankment as if in gratitude. That, to me, was the true *aarti* of the river.

Rivers also mirror our minds. When they are clean, societies flourish. When they are clogged, civilizations decay. The Ganga Basin once nurtured empires, poets and saints, from Kabir in Kashi to Tulsidas on the ghats, from Ashoka's edicts in Bihar to Tagore's poems in Bengal. All drank from the same mother's lap. Our art, agriculture and architecture were all river-fed. Even the rhythm of our classical music echoes the ebb and flow of rivers, the *raaga*s shifting with monsoon and dusk, like currents turning under light.

Today, when I walk along the Yamuna banks near Delhi, I still hear the same whisper of continuity. Beneath the stench and debris, there is still life. Fish return when we clean a patch, birds follow when trees grow and children come running when they see water clear enough to reflect their faces. That reflection is not just of their features, it is of possibility.

The science of rivers is not separate from the art of living. To understand flow is to understand forgiveness. To plant on a riverbank is to believe that movement and rootedness can coexist. If the peepal taught me silence, the river taught me song. Between them lies all my education, Granny's tales, hydrological truths, ecological service and cultural belonging. When I think of it now, I realize: we never learn *from* rivers, we learn *with* them. They are companions, not teachers; reflections, not sermons. And as they flow through the land, so do they flow through memory. The Ganga, the Yamuna, the nameless village streams—all continue to whisper one lesson to those who listen: 'Flow, even when no one watches. Flow, even when the world forgets.'

Chapter 11

Silent Action for Birds and Insects

Do what is good for the insects and birds. Create habitat. Grow vegetation that feeds the squirrels and shelters the small mammals. They know best. They have had time—millions of years—to perfect the art of balance. Humans are newcomers on this planet, still fumbling, still noisy, still trying to fit in. We have not yet learned how to adjust to the natural world. Our brains run faster than our hearts, our wants louder than our needs. We are trapped in the web of our own inventions, anxieties, appetites and artificial lights.

Granny's advice, like most truths of the earth, was small in words but vast in meaning. She never lectured me on how to behave with people. 'You will learn that on your own,' she would smile. 'But learn how to behave with trees and soil first. They have older manners.' Her sentences carried the stillness of monsoon afternoons. 'Do what is good for the birds and insects,' she said. 'Create maximum shade on the soil. Do all that you can to conserve water and shade.' Those were not instructions; they were incantations. They were spells that stayed with me long after her voice had turned into memory.

As years went by, I saw how right she was. Birds were the first teachers. They arrived before any syllabus was written. They did not wait for policy clearances or budgets. They worked every dawn, building their homes with patience, knowing exactly how to fit twigs and grass into the geometry of wind. Insects followed, whispering the laws of survival through their tiny movements. The soil listened and responded.

But we humans, we complicated everything. We fenced the wind, patented seeds, polluted water and called it progress. Religion took charge of forests, and politics invaded rivers. Soon, every conversation about climate change became a shouting match on television screens. The earth, meanwhile, kept turning silently, watching its noisy guests argue over ownership. I stopped arguing. I learned that the best use of breath is in blowing dust off a sapling, not in debating on podiums. Granny's voice returned each time I stood before a dry patch of land. 'Work around hurdles,' she whispered. 'Be gentle. Be smart. Don't waste your strength on confrontation. Use it to water the roots.'

So, I learned to work silently, under the sun, in the rain, through droughts and disappointments. We did not wait for governments to act. We gathered friends, volunteers, wanderers, believers. We built small circles of effort, each one a shade circle, each one a habitat. Slowly, the insects returned, and with them the birds, and with them the small mammals who had been waiting for safety. When they came, I knew we were doing something right. This is the age of individual action, of quiet revolutions. Leadership now grows like creepers, from below, not from above. You don't need flags or slogans. You need shade, patience and persistence. The soil does not hurry, and yet everything grows from it. There is no point in discussing policies and politicians. They are temporary gardeners of temporary gardens. The real garden is underfoot. The roots know who truly cares.

When people ask me what I have learned from decades of planting, I tell them this: Listen to the insects. Follow the birds. They are never wrong. They do not divide the world into borders, faiths or ideologies. Their only creed is balance. This insight, this quiet wisdom, became our survival kit during the darkest times, when funding dried, when drought came, when criticism rained. It reminded us to keep planting, to keep shading, to keep conserving. Granny was right all along. Do what is good for the birds and insects. Everything else will fall into place.

Chapter 12

The Ghosts We Carry

Over time, something unexpected began to reveal itself. It took me years to recognize what Granny had truly given me. It wasn't merely belief in Krishna or Hanuman or ghosts. It was the philosophy that strength comes from relationships with Gods, with ancestors, with unseen forces. It was her way of saying: *You are never alone. You are held by presences, seen and unseen.* That conviction stayed with me.

Later in life, when I travelled alone across dusty roads, planting saplings in desolate fields, I still felt her presence. The road was long, the work exhausting, but I never felt abandoned. Somewhere, in the rhythm of my breath and the sway of trees, I heard her voice: *Don't fear; you have protection.* The older I grew, the more I realized that Granny's ghost stories were not just stories. They were philosophies disguised in shadows. The ghosts she spoke of were not only in trees but also within us. Beliefs can haunt more stubbornly than shadows, memories can appear more suddenly than spirits, fears can rattle louder than shutters in the wind. I began to see every belief as a kind of ghost. Some were flowering ghosts, gentle, fragrant, comforting. These were the beliefs that gave me courage: the idea that kindness matters, that truth will prevail, that love heals. I could almost see them hovering in the background, like the soft white champa blossoms that fell in our courtyard, blessing us with their scent.

When Granny swept them into a little heap, she would whisper, 'Every flower is a blessing, sweep gently.' At that age, I thought she meant only that the blossoms were delicate. But years later, when life pressed its weight on me, I realized she was training me to hold beliefs with tenderness. Granny would never allow us to laugh at a bare tree. 'It still houses souls,' she said, 'restless ones who wait until remembered.' As a boy, I hurried past those skeleton trees, afraid of their creaking shadows. But later, I realized she was teaching me about the stubborn thoughts that cling even when they have no life left in them. I carried them for years: fears that my work was meaningless, doubts that my saplings would never survive, suspicions that my own faith was too fragile to last. These were my leafless ghosts. They gave no nourishment, yet they refused to release me. And only by naming them, acknowledging them, remembering where they came from, could I let them rest.

But not all ghosts were beliefs. Some were memories, refusing to fade. They appeared at odd hours, like spirits peering from tree hollows. We carry them with us: the faces of loved ones gone, the voices of teachers long silent, even the laughter of childhood friends who vanished into the fog of years. When Granny told us that ghosts linger until remembered, I did not understand. I thought she meant rituals, incense and lamps lit for ancestors. Only later did I see she was speaking of the memory itself. Forgotten lives become restless within us. Acknowledged lives find peace. That is why remembering matters, not only in rituals but in the way we carry stories forward. Every time I plant a peepal, I feel I am remembering her. Each sapling is a kind of lamp, a gesture that says: *I have not forgotten*. And every time I tell her ghost stories to younger listeners, I am giving her restless ghosts their rest. She lives now in my speech, in my saplings, in my solitude. She is in the flowering ghosts that bless, the fruiting ghosts that warn, the leafless ghosts that beg remembrance. She is in the peepal's whisper, in the dusk wind, in the unseen companion who walks beside me still. Granny and the ghosts—the two presences inseparable in my memory. One gave me stories, the other gave me metaphors. Together, they gave me a way of seeing the world where nothing is only what it seems. Every shadow carries a lesson, every silence a presence, every ghost a gift. Ghosts, for me, do not begin in horror films or the howling of wind on abandoned roads. They begin in the courtyard of our old house, under the faint amber glow of a kerosene lamp, with Granny seated cross-legged on her *charpai*. Her sari was tightly tucked, her posture as firm as her authority, and her eyes sparkled with that peculiar mixture of mischief and command. Around her, the night spread itself like a theatre curtain, dogs barking in distant alleys, owls announcing themselves from neem trees and the air heavy with the smoke of smouldering dung cakes. We, her grandchildren, huddled close, wrapped in the warmth of both fear and faith as we waited for her ghost stories. We had no television or radio dramas to hook us. We had Granny. And Granny never needed a script.

She was a master scientist of emotion. She knew that a child's brain learns through curiosity, not commandments. Every ghost story was an experiment, part psychology, part entertainment. She spoke of flowering trees where gentle spirits played, of fruiting trees haunted by greedy ghosts who punished theft at night, of leafless trees where restless souls clung until remembered. By turning shadows into personalities, she gave our imaginations a safe playground. What we called 'fear' was only an early rehearsal for wonder. We shivered, we laughed, we dared one another to step out in the dark. Granny teased our fear, clapped when we jumped and turned dread into play. Behind the stories lay discipline. We learned not to steal, not to disrespect and not to wander recklessly.

Every ghost in her stories had a name. Even stray cows and cats in our lane were named; why should a spirit be nameless? And so, the ghost in our peepal became Baburam. 'Baburam lives in that peepal,' she declared. 'He died long ago in Kathunangal near Amritsar, but never left the trees.' From that day, Baburam became a member of the family: unseen, unpredictable, but never unwelcome.

When milk was spilt, it was Baburam. When a child tripped, it was Baburam who had pulled his leg. When someone sneezed, Granny would mutter, 'Baburam walked past.' He was everywhere, an invisible referee to our daily clumsiness. Once, a daring boy placed a roti under the tree at night. By morning, it was gone. A dog had likely eaten it, but we knew better; it was Baburam's dinner. That boy strutted like a conqueror, proud of having fed a ghost. If Buddha found enlightenment under a peepal, we found Baburam under ours.

Granny spoke in the language of ghosts because that was the tongue of her time, but her meaning was deeper: *What you cannot see can still guide you. Ancestors, memories, nature itself, these presences are real, and to forget them is arrogance. Respect them, and you walk wiser.*

So now, when I plant a peepal sapling, I smile. I know I am not just planting shade or oxygen. I am planting stories, humour, memory and myth. I am planting Baburam and Granny again in the same soil—one invisible, one eternal.

Chapter 13

Becoming 'Peepal Baba'

In India, names are never just names. They are verdicts, blessings, sometimes even curses. They cling like incense smoke, subtle but inescapable. Some are given at birth, some are stolen by friends and a few are gifted by the crowd, the only ones that truly stick. *Peepal Baba* was one such name, not invented in a meeting room, but born in the dust of village paths, among cows, children and saplings. It was not chosen by me; it was whispered by them. It began casually, almost comically. Somewhere between a prayer and a tease, someone said, '*Arey ye toh Peepal wale Baba hain!*' And that was it. No formal ceremony, no certificate, no speech. A villager's half-joking remark became an identity that outgrew my given name.

At first, I laughed it off. I was too busy digging pits and tying protective rings of thorny branches around the saplings to think about names. But the phrase travelled faster than any pamphlet could. By the time I reached the next village, someone had already heard about a man who planted peepal trees for no personal gain. Word of mouth became my press release. The radio of rural India—gossip, temple chatter and chai-shop debates—had done its job.

The name *Peepal wale Baba* carried more authority than 'environmentalist'—a word too foreign for most tongues then. 'Baba' was a term of affection, not renunciation; a recognition of faith and continuity. Over twenty years, 'Peepal wale Baba' shrank to 'Peepal Baba', just as rivers shed syllables on their way to the plains. The 'wale', that middle particle of humility, disappeared. Maybe it dissolved in the soil. Maybe the name matured, like the trees themselves. By then, my friends had shortened it to PB. A man, a tree and a movement, all reduced to two syllables: 'PB'.

The name changed me too. It became less a tag and more a trust. When villagers called out '*Peepal Baba aa rahe hain*', children ran to fetch water. Women came with brass pots. Men brought spades. The title carried an expectation—to plant, to nurture, to bless. It was no longer mine to decline.

Sometimes, I wondered if my real name even mattered anymore. *Peepal Baba* had swallowed it whole, not with pride but with purpose. The identity had fused with the intent.

In Indian folklore, names are often mirrors of one's karma. Kabir was called 'Julaha' because he wove; Valmiki means anthill because he emerged from one; Tulsidas earned his suffix by writing verses sweeter than tulsi leaves. A name given by the people is a social consecration, not a branding exercise. My own name carried the same grassroots grace. I did not earn it by design, but by repetition, each time a villager pointed to a newly shaded path and said, '*Baba ne yeh lagaya tha*.' That sentence carried more weight than any award. In the villages, a name is a story in shorthand. And this one told a tale of faith, mud and photosynthesis.

My nickname began to travel faster than my scooter. Those were the days when roads were fewer, petrol cheaper and faith stronger. Wherever my two-wheeler went, people gathered. Some curious, some amused, some reverent. They came with empty plots and full hearts. Villagers began offering me land near ponds, temples and grazing fields. They wanted green cover for their cattle, shade for their pilgrim routes and grass for their goats. I promised nothing in return except trees. No contracts, no committees, just trust and sweat. They, in turn, gave me what every planter needs: soil, water and belonging.

The panchayats would meet and declare, '*Yeh zamin Peepal Baba ke liye rakhi gayi hai.*' No stamp paper. No rent. Only belief. They saw me not as a saint but as someone who might heal the land a little, the way a doctor heals a fevered body. And since my work brought fresh grass and rainbirds, the village elders began referring to me as *devta ka aadmi*—a man sent by the Gods to appease their angry soil.

The early years were a blur of motion and mud. Volunteers came in dozens, sometimes hundreds. Some stayed for a day, others for a decade. They dug, watered, argued and learned. The first rain always sorted the wheat from the chaff—those who only wanted photographs disappeared with the drizzle, while the true ones remained to watch the roots take hold. They became my real family. No one wore uniforms or badges. Their reward was the growing green, their salary the shade of the next monsoon. We slept in temples, gurudwaras and *panchayat ghar*s. Some nights, I stayed under peepal trees to reassure villagers that the ghosts were harmless. If they could see Peepal Baba sleeping peacefully, they said, maybe the spirits had changed their habits too.

The peepal tree has always been a curious citizen of India's spiritual republic. Worshipped by women, feared by men, inhabited by ghosts and visited by Gods, it is democracy at its purest. Under it, the holy and the haunted share the same address.

I used to tell villagers, 'Yes, there are ghosts on these trees. But they are friendly ones. They guard your land.' They would laugh nervously, unsure whether I was serious. Then I would lie down under the tree, close my eyes and let the night wind do its work. When I woke up unharmed, they smiled. The ghosts had accepted me.

Somewhere in Rajasthan, a local journalist, whose name I now forget, noticed the plantations. His newspaper headline read something along the lines of: 'Ek Ladka Jo Peepal Lagata Hai, *Peepal wale Baba.*' The label stuck harder after that. Soon, Doordarshan came with cameras, then reporters from France and Germany. They filmed me planting, sweating and laughing. I had no clue why they found it extraordinary. Planting trees seemed the most ordinary thing to do. Years later, when the world started speaking about 'global warming', I realized what we had been doing all along had a global vocabulary now. The dictionary had finally caught up with the fieldwork. The villagers didn't need those words. To them, I was still the same *Peepal wale Baba*, the man who came on a two-wheeler, carrying saplings and strange hope.

There was something miraculous about how this nickname opened doors. Because of it, nurseries offered me free saplings, sometimes whole truckloads. Forest officers bent rules to help. Panchayats that never agreed on anything agreed on giving land for plantations. '*Baba ke liye do bigha zamin rakho*,' someone would say, and it was done.

Money was never abundant, but generosity was. The village ecosystem functioned on barter—some brought saplings, some water, some labour, others cooked food. The *langar* of greening was always open. No project reports. No invoices. Just shared intent. And the best part? The livestock thrived. The cows, goats and buffaloes grazed in the new patches, and the villagers began calling the plants *charagah ka ashirwad*—blessings of the grazing fields.

Then came the 1990s and 2000s. India began to accelerate: highways, malls, mobile phones. Tree planting too became a business. Companies began hiring agencies to green their campuses. Farmers were urged to plant teak, eucalyptus and poplar for quick profit.

In the north, I couldn't escape it. Panchayats requested eucalyptus plantations because the timber mills of Yamunanagar and Saharanpur were hungry. I resisted at first, but compromise is also a part of conservation. I planted their cash crops, but with a twist. Along the perimeters, I quietly slipped in peepal and neem. When the poplars were felled a decade later, those humble shade trees stood untouched, guardians of the field. The villagers smiled and said, '*Baba apney peepal laga hi gaya*.' The ghosts must have laughed too.

Each region added its own flavour to the green. In Maharashtra, peepal and neem dominated. In Punjab, sheesham and bakain joined the mix. In Haryana, karanj thrived. Where the villagers wanted fruits, we planted jamun, amla, mango and lemon. Every sapling carried a story, every species a friendship.

Soon, institutions—Rotary, Lions, Inner Wheel, Rotaract—joined. They brought banners and photographers, I brought ghosts and roots. Together, we made quite a team. And wherever we planted, the newspapers followed. '*Peepal wale Baba* again plants trees with Rotary Club,' they wrote. I smiled. The ghosts must have enjoyed the publicity. Over time, sarpanches began winning elections by showcasing the green belts we created together. They called them Gochar Vikas Yojna, livestock grazing zones, and displayed them in campaign speeches. I smiled from afar. Democracy was finally rooted in soil.

Even officials and district collectors began visiting. They didn't always understand ecology, but they understood optics, and nothing looked better than a village lined with trees. Sometimes I was invited to speak, sometimes simply photographed. Either way, the trees benefited, which was all that mattered.

There were funny moments too. They brought me a sick goat, expecting miracles. I told them the goat needed clean water, not mantras. The next morning, the goat recovered, and my reputation as a healer spread faster than before. In another place, a schoolteacher invited me for a lecture but introduced me as *Peepal Baba from Nepal.* I never corrected him; it rhymed well enough.

Social Media and the Second Birth

Years later, when social media discovered me, I didn't resist. Peepal Baba was already the name by which thousands remembered the work. Twitter hashtags, Facebook pages and Instagram handles merely extended what the villages had already authored decades ago. The digital world was only echoing the rural imagination. Some call me PB, others Baba Ji, a few still say *Peepal wale Baba.* I respond to all. After all, each name is a branch of the same tree.

Today, when I sit under a peepal tree, I often think of those who cut them down for buildings or roads. Strange things follow, I've observed: sickness, unrest, loss of peace. I don't claim supernatural causes, but patterns don't lie. Perhaps when a tree that has been home to centuries of spirits is destroyed, something in the human psyche also fractures.

Granny used to warn me: 'Never challenge the unseen. Make friends with it.' She believed ghosts had their preferences; some loved flowering trees, others fruiting ones. To keep me safe, she offered me to Hanuman. 'Let him guard you. He will keep you strong,' she said. She would wake before dawn, talk softly to Krishna and by sunrise hand me a steaming glass of chai. In that simple act of devotion, she passed protection onto me. I suppose her prayers still travel through those peepal leaves. The ghosts have kept their side of the pact. They've watched over me through countless journeys, ensuring that the only things that haunt me now are unfinished plantations.

When I hear people call out 'Peepal Baba', I sometimes pause to see who they are really addressing, the man or the mission. Maybe both. Names, like trees, outlive their planters. The day a name starts belonging to everyone, you know it has found its roots. So here I am, not a saint, not a scientist, not even a full-time environmentalist; just a man once called *Peepal wale Baba*, who kept walking with his saplings until they became forests.

Chapter 14

The Sapling Network: A Quiet Revolution on Rooftops and Balconies

Everyone said it couldn't be done. 'You can't grow peepal from human hands,' they told me. 'It grows only when God wants it.' That sentence used to make me smile, partly because it sounded so final, and partly because I was too young and had no intention of obeying anyone, not even God if he stood in the way of my saplings. So, I decided to prove them wrong. That was all. No ambition, no grand plan, no NGO. Just a stubborn will with a small trowel and a large ego that refused to accept that divinity had monopolized the peepal.

And so began what the newspapers would much later call the Peepal Conservation Project. Its beginnings were quieter, humbler and definitely less funded. No tractors, no scientists; just rooftops, buckets and mothers with old sarees cut into strips for tying saplings.

Every city, I realized, was secretly green. Not in its parks, but in its forgotten corners. Peepal, banyan, gular and pakar were sprouting from drains, temple walls, electric poles and roofs. Millions of them. Nature was working overtime; humanity was busy sweeping them away in the name of cleanliness. I used to imagine the municipal trucks carrying those uprooted saplings to landfills, a slow, tragic funeral procession of wasted forests. It broke my heart. So, I began rescuing them. If people can adopt dogs and cats, why not trees?

That's how it began, the adopt-a-sapling idea, long before it became fashionable. I would collect peepal seedlings from the walls, wrap their roots in damp cloth and carry them home on my red Speedking bicycle. Later, when the scooter came into my life, the movement received its first mechanical engine. We tied gunny bags of saplings to the sides of our scooters—old Bajaj, Priya, LML Vespa—and off we went. We looked more like vegetable vendors than environmentalists. But that was our magic. People trusted us because we looked like them.

We didn't arrive in shiny jeeps with air-conditioned arrogance. We came on two wheels, dusty, smiling and very persuasive. '*Aunty, aapke roof pe do dabbe rakh dein kya* (Aunty, can we keep two boxes on your terrace)?'

No one could say no to that tone, half innocent, half audacious. Before we knew it, many volunteers were growing peepals on their terraces. Housewives were the first ones to say yes. They had an instinctive understanding of care. After all, they had kept families alive in smaller spaces. To them, a peepal was just another child who needed attention and water twice a day. Many of these women didn't even know the botanical names of what they were growing, but they knew the feeling of watching life unfold in their care. They became gardeners without realizing it. Some even began naming the saplings, '*Ye Lakshmi hai, ye Saraswati, ye Hanuman ji ka ped hai.*'

Rooftop nurseries became the new drawing rooms. We held weekly meetings there, sitting on old cane chairs surrounded by hundreds of saplings swaying like little disciples. Someone brought poha, someone brought chai and someone's child recited a poem about trees. It was as if the plants had turned into guests, and the rooftops had become temples. Our meetings were simple, informal and occasionally hilarious. Someone's living room became the headquarters. Someone's terrace became the distribution centre. We discussed survival rates and watering schedules with the seriousness of parliamentarians, except that our refreshments were better. Children took notes in their school notebooks. We kept ledgers of saplings. One house had a blackboard listing:

Peepal: 43
Neem: 12
Banyan: 7
Jamun: 3 (under observation).

After the official talk, the real business began—laughter, gossip and another round of chai with *bhajiya*s.

When the city inspectors came to check 'illegal rooftop structures', they were confused. What law could possibly cover a terrace full of trees? We told them, 'We're not building, sir, we're growing.' They smiled and left with a sapling of their own. Bureaucracy never looked greener.

More than plants, we were cultivating relationships. Every sapling had a story attached to it. Someone's child had planted it after exams, someone else after recovering from an illness. We began to see how the act of growing something healed people.

The movement began to ripple. What started in Pune spread to other cities—Delhi, Chandigarh, Jalandhar, Allahabad. People began writing letters to us, asking for 'rooftop nursery instructions'. This was long before emails or WhatsApp. We wrote back with hand-drawn diagrams and notes in pencil. It was never about scale. It was about sincerity. Each rooftop became a nursery, each volunteer a gardener, each house a silent revolution. Over time, those little rooftop plants descended to the ground, transplanted along roadsides, canals and temple courtyards. They became forests, and the rooftops became stories.

Looking back now, I realize that the rooftop was not just a place, it was a metaphor. A meeting point between the private and the public, the domestic and the divine. It was where the sky met human hands. The Rooftop Nursery Movement was not born out of funding, policy or planning. It was born out of stubborn love, the kind that refuses to accept that something sacred can't be touched. We began with no manuals, no degrees and no approval. But we had what mattered—the courage to begin, and the company of countless hands that cared.

Chapter 15

The Sacred Tools

In every gardener's life, there comes a moment when the tools stop being just tools and become companions instead. My spade, for instance, has more stories to tell than most people I meet. Its wooden handle has the sheen of a thousand mornings; its metal edge, the scars of monsoon trenches and rocky stubborn soil. Every mark, every dent, every bit of rust, a memory.

After every day in the field, I return home with muddy hands and an even muddier soul. But before I rest, the tools must rest right. Cleaning them is not a chore; it is a ritual. The mud on a trowel is not dirt; it's a story of where you've been. I knock it off gently, rinse the blade, wipe it dry and apply a thin film of coconut oil to keep the metal from weeping with rust. The brushes used to clean them are old toothbrushes, soft enough not to hurt the steel, yet strong enough to reach where cloth cannot. Each stroke feels like tending to an old friend's wound. Granny used to say, 'The day you stop cleaning your tools, the soil will stop trusting you.' I didn't understand it then. Now I do. Clean tools are not just efficient, they are symbols of respect towards labour, towards land, towards life.

In our small shed, every tool has a name and a home. The spade leans against the north wall, the rake beside it. The hand fork, the pruning shears, the secateurs, all rest on a wooden pegboard lined with their silhouettes, so that even if one goes missing, the absence is visible like a missing tooth. Even the watering cans have their own corner, turned upside down so no mosquito makes a home inside. A toolshed is a temple, small, quiet, sacred. The air inside smells of earth, oil, wood and metal, the four elements of fieldwork. I often light an incense stick there, just as Granny lit one before starting her cooking. Both acts are forms of gratitude, one for the nourishment of the body, the other for the nourishment of the earth.

In my early years, I would throw away broken things. Then, the land taught me patience. Now, every broken handle gets sanded and replaced, every loose screw tightened, every blunt blade sharpened with care. There is something deeply satisfying in bringing a tool back to life—a quiet resurrection. I keep an old whetstone, passed down from a nursery worker in Pune. Every few weeks, I sit by the window and sharpen my pruning shears, moving them in slow, circular motions. The sound, stone against steel, is like a chant. It is meditative, rhythmic and timeless.

My gloves are more than protection; they are memory maps. The leather ones from Rajasthan still smell faintly of camel dust; the cotton pair from Delhi bear the green stains of chlorophyll. They are all sun-dried, their fingers turned outward like open palms in prayer. I have caps from every planting season hanging above them, each one sweat-stained, memory-soaked and bleached differently by the sun. People collect trophies; I collect hats. Some are torn, some frayed, some faded, but each carries the mark of a thousand saplings planted. They hang like flags of small victories.

At night, when the world quiets down, I sometimes visit the shed just to look at them. The spades stand like soldiers at ease. The gloves rest like folded wings. The watering cans gleam faintly in the moonlight. It feels as if the tools are dreaming of morning, of soil, of work yet to be done. I close the door softly. They have earned their rest.

People often ask me why I speak of tools as if they are living beings. The answer is simple: they are extensions of us. A spade digs where our hands cannot. A rake gathers what our fingers cannot. A watering can carries what our palms cannot hold. Together, we become complete. Without them, I would merely be a man staring at soil. With them, I became a gardener.

Granny used to oil her kitchen knives the same way I oil my pruning shears. 'Metal,' she said, 'is like a man; it rusts without love.' I smiled then; I smile even more now. Every tool I've owned has been a witness to growth, to decay, to renewal. They've seen the first rains, the long droughts, the laughter of volunteers, the silence of trees. When a spade breaks beyond repair, I do not throw it away. I bury it near a peepal or a neem, as a mark of gratitude. 'Return to the earth you served,' I whisper. That is the full circle of all creation, and of all work.

Chapter 16

Enid Blyton to Ruskin Bond

Before I touched the soil, I turned pages. My first forests were not of roots and trunks but of words, orderly rows of sentences standing like saplings in the sunlight of imagination.

My love for books began long before I knew what I would do with them. Granny's trunk at home was my first library. It smelled of camphor, mothballs and old paper. Inside were treasures: torn fairy-tale books, fables from around the world and a few old *Reader's Digest* issues with photographs of faraway mountains and rivers. When I opened those pages, the world opened too.

In those early years, I believed stories were alive. Every book was a living being that spoke when opened and quietened when shut. I would place a flower petal between pages as an offering, as if feeding it something from my world in gratitude for what it gave me of its own.

I read whatever I could find—folk tales, moral stories, short pieces about animals or half-broken encyclopaedias left behind by my father's army transfers. The *Children's Encyclopaedia* became my atlas of wonder. I traced rivers with my fingers, repeated mountain names aloud like prayers: Himalaya, Andes, Alps, Aravalli. The world's geography unfolded like a living organism. Rivers and mountains became my first teachers. I would copy their names in a notebook, sometimes draw them with coloured pencils and imagine walking their paths. I had not yet seen the Ganga's glimmering waters, but I already knew its tributaries, its bends, its origins in the glaciers.

Then came Enid Blyton—the magician of my school days. I must have read every word she ever wrote. Her books were my tickets to freedom, friendship and endless curiosity. The *Famous Five* were my first environmentalists without knowing it. They were always outdoors, cycling through lanes, camping by rivers, picnicking in meadows and solving mysteries among cliffs and coves. Those stories filled me with a love for open skies and fearless companionship. I envied their seaside adventures, the sandwiches they packed, the ginger beer they drank and the dogs that followed them faithfully. I had neither a dog nor a bicycle at that time, but I had imagination, and that was enough. Reading Blyton was like tasting summer.

Every book smelled of grass, salt and freedom. There was something profoundly moral about her world, without preaching. Her children were independent yet kind, brave yet humble, practical yet full of wonder. They found magic in ordinary corners: a secret passage behind a bookshelf, a hidden key in a hollow tree. Perhaps that's where my fascination for hollow trunks began.

I read Blyton late into the night under a blanket, using a torchlight that overheated and burned my fingers. I remember hiding books inside my school notebooks during math classes. Numbers could wait; adventures couldn't. By the time I turned twelve, I had devoured the *Famous Five*, *Secret Seven*, *Adventure* series, *Malory Towers* and even her books about talking animals and enchanted woods. Somewhere between her elves and explorers, I found a bridge between imagination and nature. She made me believe that the natural world was not background scenery; it was a character. The forest was alive, the wind whispered clues and the sea spoke in riddles.

As I grew older, libraries became my go-to place. In every new city my father was posted to, I sought out a library before friends. Pune, Jalandhar, Allahabad, Chandigarh, Dehradun—each place had its own treasure chest of books. The librarians began to recognize me as the boy who returned books too quickly. I read like someone planting saplings, methodically, curiously, with quiet faith. Encyclopaedias, atlases, animal guides and old National Geographic issues were my favourite companions. Through them, I travelled without tickets: to the Nile Delta, the Alps, the Amazon, the coral reefs of Australia. When I finally did visit Australia many years later, it felt like déjà vu. I had already walked those deserts and coasts a thousand times in my reading mind.

If Enid Blyton lit the match, Ruskin Bond kept the flame alive.

I discovered him on a quiet school afternoon, in a small library corner labelled 'Indian Authors'. His book *The Room on the Roof* sat there, modestly, between weightier tomes. The cover had a boy looking out over hills. Something in that boy's solitude mirrored my own. From that day, Bond became my lifelong companion. Over the next forty years, I would read and reread everything he wrote—short stories, essays, poems, diaries, anthologies. His words became the quiet background music to my own journey.

In Bond, I found the India I loved: slow, gentle, leafy, human. His hills were not tourist spots; they were living beings. His people were not heroes; they were kind strangers, small children, old shopkeepers, watchful leopards and ghosts who never harmed anyone. He made kindness fashionable. Reading *Rain in the Mountains* or *Our Trees Still Grow in Dehra* felt like meeting an old friend on a misty road. He never exaggerated nature; he simply described it until it became sacred. Bond didn't use the language of revolution; he used the language of stillness. That, perhaps, shaped my own voice forever. I often quote him in my field diaries. Once, while planting trees near Mussoorie, I wrote: 'Planted peepal on the slopes today. Remembered Bond's line, "Once you have lived with the mountains, you belong to them." Perhaps I now belong to them too.'

Bond taught me to respect the rhythm of life, the poetry of rain and the wisdom of small things. Through him, I learned that a life doesn't have to be large to be meaningful; it simply has to be rooted.

As I entered my college years, my reading began to widen. I met Thoreau, Emerson, Rachel Carson, John Muir and Darwin—voices from distant lands who spoke the same language as our forests. Thoreau taught me that solitude is sacred. Emerson whispered that nature is the visible face of the divine. Rachel Carson showed me that science and sensitivity can coexist. Muir's letters from Yosemite read like psalms. Darwin's voyages taught me patience and observation.

Each of them gave me tools: philosophy, ecology, curiosity and humility. And yet, whenever their words became too foreign or heavy, I returned to Bond; his hills were home.

In my later years, I rediscovered Indian writers who had long been hidden behind colonial syllabuses: Jim Corbett, Salim Ali, Tagore, R.K. Narayan and Kalidasa. Each revealed a different face of nature—Corbett's courage, Ali's precision, Tagore's spirituality, Narayan's humour and Kalidasa's cosmic lyricism. Tagore's *Geetanjali* spoke of the divine in dew drops. Kalidasa's *Meghaduta* taught me to see clouds as messengers. R.K. Narayan's *Swami and Friends* reminded me that childhood and simplicity are not to be outgrown; they are to be tended like saplings.

When I look back now, I realize that my reading life was an apprenticeship for the life of planting. Each book was a species. Each author, a gardener.

Chapter 17

Peepal to People

No man is an island. My networks did not grow in offices or conferences; they sprouted in tea shops, dhabas, roadside shops, garden benches, under the village trees, cycle repair shops, bakeries and paan shops. In Allahabad, where I studied history, political science and literature, I spent more on books than on food. But I spent countless hours in tea shops. Those were India's true universities: unregistered, unstructured and unstoppable. At those tin-roofed counters, people of every class, caste and curiosity gathered. Tea shops became volunteer hubs, classrooms of conversation and places where strangers turned comrades in planting.

The dhabas of earlier years had personality: old men puffing on hookahs, parathas sizzling on iron *tawa*s, mustard oil rising like incense. Every dhaba had its own fingerprint of flavour. Today, highway food has lost that individuality; one sauce fits all. But in those days, dhabas were part of my green journeys. The *fauji* dhabas, especially, with their smoky parathas and generous spirits, gave me both calories and courage. Planting by highways meant the highway became my home, and food became my fuel.

I learned to keep a low profile. Talking too much invites predators. Silence is safer. When the work is done in silence, it creates its own noise years later. My method was simple: dig pits, mark the moist edges of highways, return with saplings, water and manure them, repeat. Some days, I dug in the morning and planted in the evening. Some weeks, it was a four-day cycle. With friends, the rhythm quickened, but the spade still taught patience.

I planted because it gave me joy—the kind of satisfaction butter-jam toast gives a hungry traveller on a long highway ride. My life's philosophy has always been butter-jam simple: do what gives joy, share what sustains. Tea and toast have been my silent allies. You may not find gourmet food on India's highways, but you will always find chai, poured from blackened kettles, sweet as optimism. That fuel has carried me farther than any elaborate diet plan. Grapes and bananas may have come and gone, but tea remained the true elixir.

So, when I say 'Peepal to People', it is not a slogan. It is the truth of my life. From one tree to one human, from one human to one community and from communities to a country, the connection is organic. Roots connect silently underground; humans connect in chai shops, in dhabas, in nurseries and in clusters of trust. This is not theory; this is live fieldwork.

Osho did not build temples; he built meeting places for friends, where the idea mattered more than the idol, where the gathering itself was the prayer. His presence shaped my understanding that communities, whether of monks or gardeners, are not for promotion, but for projects. That became my grammar of work. But long before Osho, my Granny had initiated me into the community of the ghosts of the peepal trees. '*Till the ghosts reside on the peepal*,' she warned, '*you will continue to work. They will be your caretakers.*' When the forest spirits are with you, why fear anything?

'*Patta hi Parmatma hai*,' she used to say—every leaf is God. That became my foundation. If every leaf is divine, then planting a tree is building a temple. The peepal is not just a tree; it is the personification of God, surrounded by a parliament of ghosts who can offer counsel whenever I want it.

As a child, I was fed a steady diet of ghost stories—on hill stations, at the ghats of Benares, Rishikesh and Haridwar. Later, when newspapers were filled with UFO sightings and alien theories, I smiled. Ghosts, aliens, ancestors—all part of the same mystery. For me, the peepal was both a telescope and a temple.

I was told that peepal trees could not be grown by human hands. Perhaps that was the final push I needed.

At eleven, I decided to prove everyone wrong. Ficus plants grew stubbornly out of cracks in walls, drains and broken steps. My nursery was there for the taking, on every city corner. I knew the municipal sweeper would remove them as waste. To me, each was a forest waiting to happen. Millions of future trees were being lost to the broom. So, I began rescuing them, one bag at a time, on my red cycle, on foot and later on scooters.

I had no formal knowledge of forestry. My education was observation. Granny taught me at bus stations, in cinema halls, on train benches: 'Observe, and one day you will be wise.' Science later confirmed what she knew intuitively: ficus are keystone species, carrying forests on their branches. Aspen trees clone themselves; every cell carries an instruction manual, and even human mothers carry fragments of their children's cells forever. Life is one long nursery.

Critics came too, of course. They always do. People love to advise without knowing. They carry rituals into gardening and prejudices into forestry. I learned to smile, nod and keep working. I had to be indifferent to be different.

Movies saved me, too. A cinema hall was my temple of escape, a place to laugh, cry, dream and recharge. My life has always been filmy, but then, what is a film if not a sequence of planted stories? Through it all, balance mattered. The feminine energy of mothers, sisters and women volunteers gave my work its nurturing power. Men could dig and sweat, but women parented the saplings. Together, we created forests.

In the end, I see it all as one continuum: Osho's circle of friends, Granny's ghosts, the peepal's canopy, the critics and the crowds, the cinema halls and the nurseries, the

DNA and the dust. It is all one network of observation and trust—Peepal to People.

Railway stalls gave me books. *My Experiments with Truth* was my first. Then came Vivekananda, Raja Ram Mohan Roy, Arya Samaj, nationalist movements and the comics of Vikram and Betal. Granny had only one superhero: Mahatma Gandhi. She folded her hands every time she said his name.

From her, I learned that simplicity is born of authenticity, and authenticity comes only with honesty and integrity. 'You will be as good as the books you read,' she would say. And I believed her. Simplicity became my inheritance. Pocket money was meagre, bicycles were my cavalry and jeans became our movement's uniform. A khadi kurta, a shovel tied to the cycle, a *gamchha* around the neck, cotton bags carrying saplings—this was our fashion, our philosophy, our signature.

What started as a personal choice became a contagion. Friends and volunteers adopted the same uniform of simplicity. Denim jeans that outlived five years became bags and pouches for volunteers. Gandhi's message had passed from Granny to me, and from me to my comrades. Minimalism was not deprivation; it was power. It made me approachable, trustworthy and efficient. People helped because they saw their contributions weren't wasted on luxury. A cycle saved rent, a frugal life saved stress and a simple kurta-jeans uniform became a statement more intimidating than any flamboyant show. The irony of nature is that simplicity is the most complex law. Once you live it, doors open.

Railways taught me movement. Books taught me thought. Granny taught me simplicity. Together, they made tree planting possible, not as a hobby, but as a way of life.

Reliving the past is a pleasure. The moment a pen touches paper, forgotten doors creak open and memory returns—faces, incidents, even the smell of wet soil on a monsoon morning. Writing, I realize, is not about creating the past but uncovering it.

It is easy to teach when you have done the work yourself for decades. In the last twenty years, I have moved more into training, skilling, mentoring and upgrading. Teams, volunteers, communities near plantation sites and even employees from corporate partners all needed guidance, and I enjoyed being a teacher again. Social media became another classroom, a way to whisper into the public ear: 'Grow something.'

When I began, I never imagined running an NGO. The idea frightened me. Paperwork, committees, bank accounts! I resisted for almost a decade. Only after I met John Abraham and Mike Pandey[1] did I give in. Their words nudged me. The time had come to build a structure. And so, reluctantly, I entered the world of NGOs.

But deep inside, I was never a 'me'. When it came to sapling work, I was always a 'we'. I rarely worked alone. Friends, partners, volunteers, saplings always had company, and so did I. That is why I prefer the phrase *sapling work* over *tree planting*. We don't plant trees; we plant possibilities,

---

1 Peepal Baba, Facebook, https://www.facebook.com/peepalbaba/posts/ten-years-ago-from-archives-the-first-pair-of-shoes-i-received-from-john-abraham/2869568769803640/, accessed on 27 March 2026.

small beings who take years to become what we call trees. Our task was always with the babies.

Pens and pencils have been my allies. When pens ran out of ink, pencils carried my thoughts forward. Faber-Castell, Natraj and Apsara were my companions. Paper too played its part. I scribbled on envelopes, old bills, the back of documents. One idea, jotted on a telephone bill, became the nursery plan for a major plantation drive. For years, I found it hard to open a brand-new diary; blank pages intimidated me. Only later did I allow myself that indulgence. Still, Granny's rule stayed: *Use what you have, not what you desire.*

Books shaped me as much as saplings did. They took away months, but gave back vision. I have yet to meet a reader who is not a leader. Granny's voice still echoes: *'Think like a teacher. Work so that others can copy, repeat, multiply. If your method cannot be taught, it will die with you.'*

That is why *Give Me Trees* Trust became a common man's movement. From seed collection to nurseries, from carrying cow dung in jute bags to fixing saplings with simple hand tools, everything was replicable, teachable, doable. No machinery others couldn't afford. No theory that couldn't be tested in a backyard. That was the secret: grassroots composting of effort, where small inputs grew into forests of faith.

Later, YouTube became my classroom without walls. My first upload was simple: a volunteer holding a phone, my kurta dusty, a cycle leaning on a neem tree. I spoke about planting on a roadside patch. I didn't expect anyone to watch. A week later, a message arrived from a college student in Gujarat: 'Sir, I tried what you said. My peepal is alive.' That was enough proof that the digital forest had begun. Through all of this, I learned to keep notes. Writing down thoughts gave shape to inner clutter. On paper, ideas became action. The ones that mattered stayed; the rest dissolved. Writing became my way to sort the noise from the truth. Reading helps. Writing helps. Fieldwork helps. Together, they form a trinity of learning: imagine first, note next, test in soil. That is the rhythm of my life. And in the end, I realize that I am still obeying Granny's advice: *Keep it simple, keep it teachable and keep it alive.*

That is what it means to live, from Peepal to People.

Chapter 18

First Clean, Then Green

When one works with attention, there is no time left for self-pity. Hands move, the mind steadies, the heart forgets to complain. I began to see that work, even the most ordinary, was a form of prayer performed with both hands—a liturgy of action. Later, when I found myself bending over saplings, digging soil, carrying water cans, even writing late into the night, it was this same understanding that accompanied me: meaning gathered wherever hands move sincerely.

In those years, my companions were not sages or saints but two mischievous cats named Heer and Ranjha. Their world was small: a cushion here, a window ledge there, yet they carried themselves as though they were emperors of wisdom. They knew instinctively what herbs to nibble when ill, what corners to seek when wounded, what foods to reject without fuss. Watching them, I realized that animals never attended Ayurveda courses, never leafed through fat volumes of pharmacopoeia and yet they survived with an elegance we humans—restless and overeducated—could not imitate. Heer curled into stillness, Ranjha leapt with curiosity; together they became my reminders that life was not meant to be conquered but observed. They were living parables of the forest itself—self-healing, self-sustaining, asking only to be left alone.

We humans, anxious to control, often forget that forests do not need us; it is we who need them. Forests plant themselves. The breeze and the bird carry seeds farther than any of our schemes. Our true role is not mastery but protection, not command but humility.

From this truth grew another that I often muttered half in jest, half in earnest: *first clean, then green.* Humankind, with its brilliance for invention, also carries a talent for toxifying. We pollute even our solutions: wind turbines that need mountains of concrete, electric batteries that leave behind poisonous scars. Before rushing to add more, the wiser act is often to remove, to cleanse, to allow the soil and water to breathe. I saw it in the smallest acts: drains clogged with plastic where volunteers wished to plant trees, fields littered with debris where eager hands wanted to scatter seeds. I would sigh, laugh and remind them: *hum toh safai karamchari hain*; sweepers first, gardeners later. Only when the ground is cleared can the green take root.

At home, a few pots became a nursery, and the nursery became the sapling network. People who initially mocked the idea returned to say softly, 'We will water it, don't worry.'

Something ancient stirred in them when they greened the Earth. Reverence, caution, longing—they could not explain why, but they felt it. That was the real victory—not the sapling in soil but the shift in a heart. Doubt, when tended gently, often blossoms into care.

Even as saplings multiplied, so did pages. Late nights found me bent over scribbled diaries, transcribing half-legible lines into typewritten sheets. Eventually, in a hospital bed, I balanced a laptop on my knees while tubes snaked around me. Nurses teased me, calling it my second office. The first hundred pages took six months, the next hundred barely two. Momentum, once found, is its own companion. Writing, I discovered, was another form of gardening. You prune sentences, rearrange paragraphs, let light fall where density has grown too thick. With each night's work, the voice inside grew steadier. Consistency brought rhythm, rhythm gave voice and the voice carried me forward.

Chapter 19

How Many?

People ask me this simple question, whose answer is very complicated. 'How many trees have you planted?' they ask, expecting a neat number, a figure that fits into a headline or a record book. But for me, numbers were never the point. I never counted shade in digits. I counted it in sighs of relief, in the hum of insects, in the flutter of wings. I knew that once shade appeared, insects would follow, then birds, and slowly, a habitat would begin to breathe again. For me, saplings were not numbers; they were relationships.

Nevertheless, forty-eight years of work do add up to something. You cannot move through four decades of soil and seasons without leaving behind a forest. But even that forest is not mine alone. I have never worked alone. My journey has always been shared, with friends, volunteers, workers, groundforce, donors and those nameless people who stopped by just to pour a bucket of water on a struggling sapling. Each one of them is a root in this story.

What began as a solitary act of planting soon became contagious. It went viral long before social media existed.

And that, perhaps, is my greatest joy. Because the idea was never about *me* doing it, it was about everyone doing it. My role was to teach, to create living models, to turn barren land into classrooms of biodiversity. These were our open laboratories: visible, breathable, learnable. People came from everywhere to see, to learn, to participate and then they went back to create their own green revolutions. We celebrate imitation. When someone copies us, we rejoice. That was the whole idea—to make replication easier than invention. To turn planting into a shared grammar of hope.

Over the years, we have trained students, homemakers, farmers, panchayat workers and schoolchildren, anyone who was willing to hold a sapling with love. We have spent more time teaching than planting, and that is something I am proud of.

If I must put a number to it, perhaps to satisfy those who live by data, then yes, the count runs into millions. But for me, the true count lies in the number of hearts changed, hands dirtied and eyes opened to the miracle of growth.

When I look back, I realize how primitive our beginnings were. There was no internet, no Google, no GPS, no WhatsApp, no YouTube tutorials to tell us how to graft, mulch or irrigate. Logistics were uncertain, communication was by post and fundraising meant knocking on doors. Even transportation was a gamble—rickety scooters, second-hand jeeps, borrowed trucks and long walks under the sun. Yet somehow, it worked. Because nature doesn't wait for technology, it rewards persistence.

Today, when I see our volunteers planting with drones, mapping with satellites and connecting on global platforms, I smile. We began when none of that existed, yet we created something living, something that continues to breathe and expand on its own. So, when people ask, 'How many?', I answer, 'As many as it takes to give the Earth a second chance.'

Chapter 20

Planted Other Species Out of Guilt

At first, I planted other species out of guilt. Yes, guilt. The kind that quietly follows you after every sapling you plant, when someone comes and says, 'Arrey Baba, don't plant peepal! It will break the walls, it will crack the house, it will bring ghosts.' And so, out of guilt, I began to plant neem.

Neem had no enemies. No one objected to it. It was safe. I could plant neem anywhere, and people would nod approvingly. I found comfort in its acceptance. It became my camouflage, my socially acceptable green companion. But then came the next wave of advice. 'You should plant sheesham,' people said. 'It's hardwood, it has value. The world respects timber, not shade.' So I planted sheesham. Then came another chorus: 'Fruit trees, Baba! People need food, not ghosts!' So I planted jamun—generous, purple and patient. And once you start planting fruit trees, you cannot stop. Soon came mulberry and mango. Bananas and guavas followed, wherever soil and space allowed.

People are people. They love to instruct. Some request politely; others order. And we, planters by mission, not by argument, agreed, because all we needed was permission and space. The rest we could negotiate with the soil. Behind every neem, jamun or mango I planted, a silent peepal was waiting for its turn. I would nod, smile, and say, 'Yes, yes, fruit trees,' while quietly slipping in a peepal sapling somewhere in the corner, near a drain, a temple wall or the edge of a panchayat pond. That was the real mission—to find a home for the peepal.

In Maharashtra, people allowed me to plant gular. 'It is sacred,' they said. Sacred was a word that opened doors, and I entered with my nursery pots like a pilgrim with offerings. In Marathwada, people adored neem, so I filled their lanes with neem. In Rajasthan and Gujarat, they wanted hardy species from the Aravalis: lasora, chamrod, baheda, nirgundi, vajradanti, henna and karonda. Medicinal, drought-tolerant, proud little species that could stand in heat where even hope would wilt.

We began sourcing saplings from the Aravalis, transporting them hundreds of kilometres, and planted lakhs of them in the Delhi-NCR region. Every state, every district had its own temperament, its own favourite species. India is like that, a continent of opinions, and I learnt to listen.

Starting with the peepal, I ended up planting more than a hundred species, maybe more. It was not a plan; it was a slow unfolding, a theatre in which every tree played a part, and I merely followed the script written by the forest itself. I had set out to plant peepal. Only peepal. But the peepal maharaj had other designs. He was using me, using my guilt, my curiosity, my need to please, to plant forests. To multiply life in ways I could not have imagined. Through our nurseries and community work, peepal made sure I also planted his cousins, neighbours and extended family of figs and forest species.

Over time, I realized: this was not guilt anymore. It was grace in disguise. What began as one man's apology for planting a forbidden tree became a nationwide orchestra of green, a chorus of roots, a carnival of species. And I, the accidental conductor, simply kept waving my spade like a baton.

Chapter 21

Neem, the Monk Tree

Some say the neem is native to India; others claim African origins. I have no patience for such debates. The neem has long made itself at home here, and that is what matters. Villagers still point to neem trees two or even four centuries old, standing like stoic sentinels, proof that, whether guest or native, neem belongs.

Neem was always accepted by all people, villages, towns and cities. Villagers in Gujarat and Madhya Pradesh still trace family histories through neem. 'This one was planted when my grandfather was married,' they say, touching the trunk like a family heirloom. Two centuries of continuity, shade after shade. No one ever said no to neem. It turned out to be the most acceptable plant of all, a tree that belonged everywhere, to everyone, without argument or objection. In a country divided by language, caste, climate and soil, neem became the quiet symbol of national agreement. It was the practical fellow among trees—low-maintenance, steady, bitter in taste but generous in shade. Wherever I went with peepal saplings, there were objections: 'Too big, Baba,' 'It breaks the wall', 'Ghosts sit on it.' Some agreed to jamun, others to sheesham, but no one objected to neem. Not once.

My own journey with neem first began in Pune. The first sapling stood near our bungalow on Range Hills Road inside the Kirkee Military Station. That first batch of neem trees looked like schoolchildren in uniform. Soon, the nursery became a movement. In the early years alone, we planted 35,000 saplings across Pune. Over time, the number crossed two lakh. Every cantonment, school and college got its share of neem. Where other trees demanded care, neem only asked for a little space and a little patience. And so, when the *Give Me Trees* campaign took shape, neem quietly rose to become our second-highest planted species, the working-class hero of our movement. Forest nurseries gifted us other species—mahogany, karanj, tamarind, jungle jalebi, sheesham, goolar—but neem was my constant companion. By the 1990s, half of Maharashtra's districts had green footprints: Pune, Solapur, Satara, Kolhapur, Karad, Tuljapur, Beed, Osmanabad, Jalna, Aurangabad and Mumbai. Each site told its own story of survival and local love.

When I began travelling for neem plantations, the country unfolded before me as a vast canvas of dusty roads, sleepy villages, stubborn soils and eager children. And everywhere, like an old acquaintance, neem was waiting, sometimes as a half-forgotten stump in a panchayat yard, sometimes as a lone soldier guarding a canal, sometimes shading a bus stop where tin shelters had rusted away. Neem always seemed to know where it was needed most.

Highways were my first battleground. Picture Rajasthan in May: a burning tar road where the sun fries not just skin but optimism. Trucks roar like angry elephants. The roadside soil is more gravel than earth. 'Nothing will grow here,' villagers said, shaking their heads. But neem is a contrarian. It sends its taproot drilling down through hostile strata, finds the hint of sub-soil moisture and declares: '*Bas, ab main hoon yahan.*'

Within a few years, those same roads were flanked by rows of neem: patient traffic policemen, green sentinels offering shade to travellers who once carried wet cloths on their heads. 'He who plants a tree, plants hope,' says an old

African proverb. On those highways, neem was hope: bitter in taste, long in shade.

Along irrigation canals in Punjab, neem revealed another face—that of an engineer. The sandy embankments often collapsed after repeated wetting and drying, forcing farmers to rebuild them endlessly. But neem roots stitched the soil together like a grandmother sewing torn clothes. Farmers leaned on their lathis and said, '*Neem ke jhad se kinara toot-ta nahi.*' No soil mechanics textbook could have explained sheer strength so poetically.

Bus stops gave neem its comic moment. In one Maharashtra village, the government installed a tin rain shelter with much ceremony. Within a year, it collapsed, eaten by rust, its sheets now fencing goats. But beside it, a neem sapling we had planted stood tall, spreading green arms. Passengers laughed, 'Tin is MLA ka promise; neem is God's promise.' Bitter humour, but true.

In panchayat grounds, neem took on gravitas. The microclimate itself seemed to discipline tempers. Ten degrees cooler under its canopy, arguments mellowed, egos softened. Kabir had it right: '*Kadva nimb satya samaan, kasak toh hai, par davai bhi wahi hai.*' Bitter neem is like truth; it stings, but it heals.

The village pharmacy followed us everywhere. Every part of it is useful. Neem leaves in rice bins against weevils, twigs chewed for dental care and toothaches and smoke rising from courtyards to chase mosquitoes. Leaf paste for rashes. Neem oil to get rid of hair lice. Nothing wasted, nothing idle. A farmer in Bihar laughed, 'Neem is our government hospital: always open, no strike, no bill.' When children resisted chewing neem leaves during Holi, grandmothers bribed them with jaggery. Bitterness turned into a sweet ritual.

Modernity tries to repackage what we already know. Shampoos boast 'neem extract'; creams flaunt 'anti-bacterial neem essence'. I smile. My grandmother did all that decades ago, boiling neem leaves in a bucket, pouring them on her hair every Sunday, her scalp fragrant with health and humility.

Travelling South, neem changed its accent but not its character. In Tamil Nadu, it was *Vembu*, offered to Mariamman during summer festivals. In Andhra, *Vepa* was guardian of fields. In Karnataka, *Bevu-Bella*, neem and jaggery, were the new year's reminder that life is both bitter and sweet. In Maharashtra, *Nimb* was tucked into kitchens and cattle sheds. Everywhere I went, neem spoke the same language of endurance, even if its name changed with geography.

And biodiversity seemed to travel with it. In Uttar Pradesh, mynas feasted on neem fruits. In Madhya Pradesh, bees buzzed around flowering neem in April. In Bengal, villagers told me that neem groves invited owls, guardians against rats. Scientists called it 'habitat service'. Villagers simply called it 'blessing'.

Shade was neem's ultimate currency. 'What man does under a tree, God does not see', goes a Spanish proverb. Under neem shade, men rested, argued, loved, plotted, repented. I once napped under a neem by a canal and woke up to find goats chewing my sleeve. The herder laughed, 'Goats like humans more than neem.' Maybe, but neem still gave us shelter. Through it all, my rhythm stayed constant: two dozen saplings per site, never less. It wasn't superstition, only the number I could personally tend with care. Twenty-four neem babies at a time, a manageable promise to the Earth.

The Beauty in Bitterness

When science caught up with what grandmothers had always known, we learnt that Neem contains *azadirachtin*, *nimbin* and *nimbidin*—natural compounds that make it antibacterial, antifungal, antiviral and insecticidal. Long before laboratories proved it, neem leaves had been our first line of defence. Azadirachtin disrupts insect life cycles, a natural pesticide without collateral damage; nimbidin fights inflammation; nimbin heals skin and wounds. Every molecule in neem seems to hum with purpose. Ecologically, it is a quiet warrior. Its deep roots bind soil and resist erosion. Its leaf litter enriches humus, its canopy reduces ground temperature and its fallen twigs become carbon recyclers. It endures salt, drought and neglect—the virtues of a saint disguised as a tree.

The Monk Tree

If peepal is philosophy, neem is pragmatism. If mango is romance, neem is resilience. Bitter, yes, but not cruelly so. Its bitterness is the taste of medicine, truth and survival. Even ghosts, villagers told me, behave politely under neem. '*Neem ke bhoot bhi doctor hain*,' someone once joked. I laughed. Perhaps even the supernatural respects its antiseptic temperament. To me, neem is the monk among trees—barefoot, self-sufficient and radiating calm through austerity. It asks for little and gives much. It is the silent teacher of patience, the living embodiment of sustainability.

It shows that true strength lies not in speed or glamour, but in endurance.

Today, as I travel across India, I still find neem waiting, along highways, canals, panchayats and schools, standing like old judges, shaking their leafy heads at our foolishness yet forgiving us with shade.

Sometimes I plant a new one, knowing I may not live to see it mature. But someone will. And when they rest beneath its branches, wiping sweat from their brow, they will unknowingly bless the nameless planter. That is neem's magic: You plant bitterness, and out of it grows the sweetest blessing of all—survival.

Chapter 22

Pedon Ka Ashram

After years of working with the land, certain patterns began to feel almost sacred. Ask. Trust. Take help—that is the sacred sequence of all life. A seed asks the soil for permission to stay. The soil trusts the rain to come. The rain takes help from clouds to fall. Nothing grows alone. That has been my first and last lesson in ecology, and in life.

Skill, stamina and steady nerves are not qualities you learn in schools or seminars. You learn them from the land itself. From planting in the scorching May heat, from watching your saplings wither and coming back the next morning anyway, from standing in the rain to guarding them from cattle and wind. No university teaches this. The soil does.

The modern world is obsessed with making life in laboratories. Synthetic DNA. Genetically enhanced humans. Crops that do not need pollinators. Chickens that never see the sun. Yet, in the rock pools we dig on our sites to collect rainwater, I see the real miracle—simple, ancient, unedited by man. Within days, those shallow puddles become entire worlds. Frogs appear from nowhere. Algae paint the stones green. A dragonfly lands, lays eggs and vanishes. Creation rehearses itself quietly, without any patent. Sea becomes land and land becomes sea. That is the magic of gravity. Every grain of soil has once been ocean floor, and every droplet in the air was once a river. We are not separate from any of it. When I look at a puddle after rain, I do not see mud. I see memories. The memory of tides, monsoons and migrations. The entire Earth is in rehearsal for the next act of life.

Observation has been my first guru. Before I learned any textbook science, I learned to watch. Looking out of a window was my first education. I watched light shift through the day, how sparrows changed direction with the wind, how trees predicted rain before the sky did. Granny would often catch me staring at the same tree for hours and ask, 'What do you see?' And I would say, 'Everything is moving, even what looks still.'

That habit never left me. Years later, when I travelled across the country, I found that observation was the bridge between experience and wisdom. Looking is science. Seeing is spirituality. And both are needed if one is to understand nature.

We cannot have one agriculture for all ecosystems. India alone has a hundred types of soils and climates, from the salty sands of Kutch to the red earth of Chhattisgarh, from the terraced fields of Himachal to the black cotton soil of Maharashtra. Each demands its own rhythm, its own patience. Yet the modern world wants a single formula, a uniform seed, a uniform market, a uniform taste. One Earth, one agriculture, one corporation. It sounds efficient, but it is violent.

Corporations are buying land the way empires once bought kingdoms. Not to grow food, but to control it. They will decide what we eat, how it is grown and how it is priced. When food becomes a product, hunger becomes a market. When agriculture turns into industry, it loses its soul. This is the agriculture of violence, chemical, economics and moral. The soil is no longer mother; she becomes machinery.

But nature has her own quiet rebellion. Weeds grow where herbicides fail. Indigenous seeds survive floods that kill hybrids. A peepal sapling cracks concrete and returns the world to its original state, alive.

Life means we are part of one food web. Every cell in our body is borrowed. Every breath is recycled. The leaf that gives oxygen today will fall tomorrow, decay and become soil for another leaf. Nothing is wasted. Even death is a form of nourishment.

Once you understand a forest, you understand everything—politics, economics, architecture, even theology. A forest is not just a collection of trees. It is a constitution of coexistence. Every species has rights and responsibilities. No one dominates, and yet everyone thrives. The smallest termite and the mightiest banyan both serve the same law: give and receive.

The Birth of Pedon ka Ashram

From this understanding grew the idea of Pedon ka Ashram. The word *pedon*—trees—comes from the Sanskrit root *pād*, meaning both 'foot' and 'foundation'. An ashram, in the truest sense, is not a retreat from the world but a return to its essence. So, I imagined an ashram where soil itself would be the teacher.

It began as a simple experiment. Small community nurseries, open-air classrooms and shaded spaces where children could learn to plant instead of merely studying diagrams of plants. But slowly, it became a movement. Farmers, volunteers, students, homemakers, soldiers, scientists, all began to gather. They came not to pray, but to participate. At Pedon ka Ashram, every spade of soil was treated as sacred. Every sapling was a student. People learned not just about planting but about patience, about failure, about waiting for roots to take hold. We did not have sermons. We had seasons. We did not have lectures. We had compost pits.

It caught the imagination of the common man because it was simple and truthful. At a time when people were tired of noise and slogans, here was a place where silence itself was an education.

And the soil never disappointed us. It gave more lessons than any scripture.

Just how much Pedon ka Ashram resonated with other people only struck home in 2005, when Doordarshan, the same channel that had shaped our childhoods with Krishi Darshan and news bulletins, aired a ten-minute short film on our peepal plantations—my first televised moment! Newspapers had been following our work for years, but television carried it like the wind. Suddenly, people across towns and villages saw what we had been quietly doing for decades, planting faith in the soil.

It was not fame that mattered that day. It was recognition of the invisible. The camera captured the rustle of leaves, the sweat on the brows of volunteers and the sight of children carrying saplings. The visual medium has a strange power; it doesn't just inform, it infects people with inspiration. That day, thousands called to say, 'We want to do this too.' That was the day Pedon ka Ashram stopped being mine. It became everyone's.

Over the years, I have realized that the same laws govern both ecology and emotion. What erodes soil also erodes the soul: greed, ignorance and arrogance. What restores the land also restores the heart: care, patience and compassion. We have built rock pools on our sites to store monsoon water, but they have become mirrors too. When you lean over them, you see not your face but your reflection in nature, a reminder that we are not masters of the Earth, only temporary guests. Sea becomes land, and land becomes sea. Empires fall. Languages fade. But the forest remains faithful to the Earth.

A Living Ashram

Pedon ka Ashram is not a building. It is a way of seeing. It can exist in a field, on a rooftop, in a school compound or in the mind of a child holding his first sapling. It is a space where the disciplines of soil, science and spirit meet. Where a gardener becomes a monk, and a monk becomes a gardener.

We live in a time of synthetic life and digital distraction. But even as technology runs ahead, the rhythm of the planet remains ancient. Gravity still pulls, rain still falls and a peepal leaf still trembles in the wind for reasons science can explain but poetry alone can feel.

Observation remains the oldest technology of all. Look long enough at a tree, and it will tell you the story of evolution, economy and ethics. All at once. And that, to me, is the true purpose of Pedon ka Ashram—to remind us that every handful of soil contains both history and hope.

Chapter 23

The Amazing Peepal Tree

If there is a cathedral of photosynthesis, it stands quietly in every Indian village: a peepal tree with roots like folded prayers and leaves that tremble even in still air. My childhood classroom was in its shade. While others prayed to it, I listened. Its rustling was not divine mystery; it was respiration—the subtle rhythm of exchange between leaf and sky, soil and soul.

Granny's stories had already led me to believe this was no ordinary tree. Decades later, science confirmed what faith had intuited: the peepal works while the world sleeps. Unlike other trees, it continues to absorb carbon dioxide during the night in addition to its regular daytime photosynthesis. In simpler words, the peepal doesn't rest.

People often come to me with questions that sound scientific but are really spiritual in disguise. One of the most common among them is: 'Baba, is it true that the peepal tree gives oxygen at night?' And I smile every time I hear it, because behind that question is not ignorance, it is awe. Awe that something so silent could be so alive. But as both a planter and a lifelong student of nature, I have learnt that truth is often a combination of fact and faith. So let me tell you the full story of what really happens inside a peepal leaf when the sun sets.

During the day, every leaf of a peepal tree is a miniature solar panel. The green pigment, chlorophyll, captures sunlight and turns it into energy through the process we call photosynthesis. The reaction is beautifully simple: sunlight, water and carbon dioxide come together to form glucose and oxygen. Each peepal leaf becomes a tiny factory of oxygen, feeding life into the atmosphere, and cleaning what we exhale. Under its broad canopy, I have often rested after long planting days, feeling that the very air beneath it carried a rhythm, a slow pulse of life invisible to the eye. When you have planted thousands of them, you begin to understand that photosynthesis is not just chemistry; it is the Earth's heartbeat.

But the sun cannot shine forever, and night arrives, quietly closing the green factories of daylight. For most plants, this means a reversal of process; photosynthesis halts, but respiration continues. Plants, like all living things, now consume oxygen and release carbon dioxide as they break down stored sugars for energy. But in the peepal, nature has engineered a special adaptation called Crassulacean Acid Metabolism or CAM. It is a beautiful solution to the challenge of conserving water while still breathing. At night, when the air is cooler and the humidity is higher, the peepal tree's stomata, tiny breathing pores, open. It absorbs carbon dioxide and stores it as malic acid within its cells and then, as the sun rises the next day, this stored carbon dioxide is released inside the leaf for photosynthesis. It is like a diligent worker who rests only to resume duty at first light. This system allows the peepal to continue a part of its metabolic cycle at night. It does not truly release oxygen in the dark, but it does something equally profound: it keeps the carbon of the Earth cycling even when the light has gone. This is the reason people have believed for centuries that it emits oxygen at night. To me, this is poetry in biochemistry.

That makes peepal scientifically extraordinary. Functioning as a keystone species in tropical ecosystems, it sustains biodiversity, stabilizes soils, influences microclimates and acts as a living oxygen factory that breathes by both day and night. These attributes have also made it sacred in the Indian subcontinent, where its cultural continuity, as a provider of shade, sanctuary and a silent teacher, stretches back thousands of years.

Its root architecture is equally remarkable—a deep taproot system that anchors it even in rocky ground, and lateral roots that explore meters of soil, finding moisture where none seems to exist. The aerial roots hanging down from branches act as living pillars, turning each old peepal into a small biosphere.

To understand a peepal, one must understand its figs—tiny, greenish-purple orbs that feed worlds. Every fig species on Earth has a unique pollinator wasp that lives and dies inside its fruit. The peepal's wasp enters through a narrow pore to lay eggs and unknowingly pollinates the flower hidden within. Without the wasp, there are no figs; without the figs, no wasp. It is an ancient covenant written in chlorophyll and wings—one of nature's most elegant symbioses. When the figs ripen, the real feast begins. Parakeets, barbets, mynas, pigeons, bulbuls, hornbills, civets, squirrels and fruit bats arrive in turn. Ecologists call such trees 'keystone resources'; their fruiting ensures continuity of food even in lean seasons. For many birds and small mammals, the peepal is a pantry that never fully closes.

Under a mature peepal, one can study entire food webs. Termites in the roots, beetles in the bark, lizards on the trunk, mynas nesting in cavities, bats in the canopy—it is a vertical city of coexistence. Its litter bed is a composting universe. Fallen leaves rich in lignin decay slowly, nourishing fungi and bacteria that enrich the soil. The microclimate under the tree remains cooler by up to 5°C compared to the surrounding open areas. Soil moisture is conserved, wind speed is reduced and the air is filled with negative ions that refresh the lungs.

No wonder people naturally gather beneath it, not only for shade, but for peace. Beneath its crown, the heart slows down, breathing steadies and conversations soften. Sitting under a peepal for meditation lowers cortisol levels, steadies alpha brain waves and induces calm. So perhaps the peepal's night-time oxygen is not a myth after all; perhaps it is a metaphor for its deeper gift: to keep us alive in ways we cannot quantify.

Across the subcontinent, the peepal tree has been the oldest public utility. It stands beside wells, ponds, temples, courtyards and crossroads. Villages were planned around them, the tree being both compass and commons. Several people have written that a single peepal can transpire up to 400 litres of water per day, absorb nearly 20 kilograms of carbon dioxide per year, release over 300 litres of oxygen daily during daylight and oxygen concentration remains slightly higher, especially after dusk. While this has yet to be scientifically measured, in my experience they have tremendous effects on local humidity and even rainfall.[2,3] Temperatures under their canopies can drop by several degrees compared to open areas. Its roots bind eroded soils, prevent landslides and recharge groundwater. In urban spaces, it absorbs pollutants, filters dust and reduces heat-island effects. These physical effects create an environment that indeed feels more oxygenated—a truth of experience if not of equation. It is part of the reason why the tree feels sacred when you sit under it at night.

In this sense, every peepal is a living piece of ecological infrastructure, offering free climate regulation and biodiversity services worth thousands of rupees annually, though the national budget never lists it. Multiply that

---

[2] Catch Foundation, NGO Working for Environment Protection, 2019, https://catchfoundation.in/blogs/best-native-trees-carbon-sequestration-india, accessed on 27 March 2026.

[3] Ashalatha, K.L., Kallare P. Arunkumar, and Malali Gowda, 'Genomic and Transcriptomic Analysis of Sacred Fig (Ficus Religiosa)', *BMC Genomics* 24 (1), 2023, https://doi.org/10.1186/s12864-023-09270-z.

by the millions we have planted, and you begin to see the invisible economy of air it creates, what I often call the Invisible GDP of Shade.

Ayurveda too has long classified the peepal as a 'Tridosha pacifier', balancing Vata, Pitta and Kapha. Its bark has antidiarrheal, skin-tightening, anti-diabetic and cholesterol-managing properties. Its latex has anti-ulcer and anti-inflammatory properties; its leaves are antibacterial and its fruits are digestive.

Peepal is a master of adaptation. It sheds its glossy leaves in summer, grows figs in pairs and lives longer than most human dynasties. I've met 900- and 1,100-year-old peepals in Haryana, still green, still gossiping with the wind. Their bark, when boiled and mixed into cattle feed, aids digestion. Their seeds ride on bird wings, sprouting in cracks and crevices, proving again that persistence beats privilege. And no, the peepal is not parasitic; it just likes living rent-free on other trees till it can stand on its own.

In our scriptures, Lord Krishna identifies himself with the peepal: 'Among trees, I am the Ashwatha.' Its roots stand for Brahma, the Creator; the trunk for Vishnu, the Preserver; and the branches for Shiva, the Destroyer—the entire cosmic department in one leafy headquarters.

If you prefer Latin, the peepal is *Ficus religiosa*, the sacred fig, the Bodhi tree, the Ashwatha. Botanists love it, believers worship it and ghosts rent it as prime real estate.

The earliest record of peepal worship dates back to the Indus Valley Civilization. By 3000 BC, people were already bowing to it. Clearly, the tree had excellent marketing long before social media influencers arrived. In Buddhist lore, Prince Siddhartha sat beneath one such tree in Gaya, meditated and emerged as the Buddha. The descendants of that tree still stand as a living and breathing heritage. The Buddha's enlightenment under the Bodhi tree is not merely a legend; it is a case study in ecological psychology.

Every region adds its twist. In some villages, the peepal (male) is married to the neem (female). In Rajasthan, the genders are reversed, proof that even trees enjoy gender fluidity long before we coined the term. Women tie red threads to its branches, seeking blessings for children, and monks meditate beneath its shade, seeking freedom from the cycle of birth altogether. Between the two, the peepal handles both fertility and enlightenment with admirable balance.

From the *Rig Veda* to the Puranas, the tree appears as Kalpavriksha—the wish-fulfilling tree. In mythology, Vishnu hid in a peepal when demons chased him; in folklore, ancestors descend its branches to bless their descendants. Even village vets recommend feeding cattle the tree's figs for better health, perhaps the ghosts moonlight as veterinarians too.

The Living Heritage

Peepal is more than a plant; it's a philosophy. It fertilises the soil, feeds the cattle, shelters the birds, cools the air and occupies the human imagination. It stands at the junction of science and superstition, ecology and eternity. Maybe that's why I've spent a lifetime under its shade, planting, protecting, pondering. The more peepals I plant, the more I see Granny's wisdom take root in me. And I smile, because perhaps she was right all along. Ghosts don't just haunt the trees; they live in our conscience, keeping us rooted, aware and a little bit playful.

Working with the sacred fig, I finally understood why it is called *the speaking tree.* It speaks to you, not in words, but in whispers that travel through silence. It communicates in ways that are not heard but felt. Spend a few quiet hours with it, and you will find yourself listening, not to the tree but to the stillness within you. In our noisy lives, there's no room for contemplation. The world keeps shouting, and the mind forgets how to listen. But the peepal tree teaches you that silence is not emptiness; it is presence. Under its canopy, thoughts begin to settle like dust after rain. The rustle of its heart-shaped leaves is not just sound; it is language. The peepal doesn't preach; it reveals. That rustling, caused by the trembling of its long, thin petioles, is nature's way of keeping the leaf surface cool and dust-free. But for me, it feels like the tree is whispering, reminding us of the delicate equilibrium that sustains life on Earth.

Each peepal has its own temperament. Some are restless, shooting upwards as if impatient for stardom. Others are calm and spreading, generous with shade. A few are stubborn, refusing to grow unless left entirely alone. Over time, I realized that they were no different from the people I met in life—each one carrying a unique rhythm, a private will. The peepal taught me patience. You cannot rush its growth. You cannot schedule its shade. You have to earn its trust. That's how life works too. You nurture, wait and sometimes just sit quietly until something finally takes root.

Many evenings, I sat under the trees I had planted as a child in Poona, watching light filter through their trembling leaves. I recalled the time spent naming the saplings after Bollywood heroines—Mumtaz, Helen, Madhuri and Hema—turning the road into a veritable walk of fame. I smiled at the thought of them standing in a row, fluttering like divas on a breezy set, still glowing with green glamour forty-seven years later, having aged gracefully just like the heroines they were named after. The breeze made them shimmer like laughter, and I often imagined they were talking among themselves. On quiet afternoons, when nobody was around, I could almost hear them sharing secrets with the wind, stories of the past, of children who grew up, of rains that failed, of seasons that healed.

Those still standing on Range Hills Road have witnessed the slow transformation of the city: military trucks giving way to scooters, scooters replaced by cars and now electric vehicles humming quietly where cows once grazed. The road has changed, but the trees haven't lost their grace.

When I visit now, I stop my scooter under Hema. Her trunk is thick, roots exposed, leaves alive with birds. I talk to her like an old friend. I tell her about the new plantations, about the volunteers who now do what I once did alone. She listens, or at least I believe she does. The air around her always feels different—cooler, calmer and forgiving.

Perhaps that is what the peepal teaches best, that growth is not about speed but about rootedness. That the truest form of communication happens in silence. That every act of nurturing, whether of soil, soul or relationship, is a slow conversation with time itself. That is why, for centuries, when people did not know about chlorophyll or CAM metabolism, they still meditated in the peace of its shade. It is why temples were built around peepal trees, why people tied threads to its trunk, circled it in prayer, lit lamps at its roots. It is why the story spread that it emits oxygen at night. It was the only way to ensure that people continued to protect it, through reverence wrapped in legend. The peepal tree may not flood the night with oxygen, but it fills it with meaning. It teaches us that life is not a straight line between day and night, light and dark, science and spirit. It is a cycle of giving, receiving, storing and releasing. If photosynthesis is the act of inhaling light, then respiration is the act of remembering it.

When I explain this to students today, in schools, colleges or through digital classrooms, I tell them that science and spirituality are not rivals. They are two languages describing the same truth. Science explains how the peepal breathes; spirituality explains why we must let it breathe. With the peepal, science and faith dissolve into the same truth: everything breathes together—humans, trees, air, and soil—in one continuous conversation of carbon and light.

Every tree I have planted has been a lesson in faith. You never know if the seed will sprout or if the sapling will survive the first summer. Yet you plant anyway. That is faith; the same faith Granny lived by when she whispered her stories under the lamplight, believing that her words would find a home in us. The peepal stands as the living proof of that faith. You water it, protect it and wait. And one day, without any announcement, it begins to speak, not with sound, but with shade. Even today, when I plant a new sapling or stand before an old peepal, I feel as though Granny is there somewhere, smiling, nodding and saying, 'See, I told you they talk.'

The fieldwork I have done all my life began here, under these speaking trees. They taught me everything: patience, humility, the art of listening and the quiet joy of doing without expecting. In their company, I learned that belief doesn't always come from temples or scriptures; sometimes it grows from the soil, one root at a time.

The peepal tree was mother and monster, both. It was worshipped in the morning and avoided at night. Women tied red threads around its trunk, whispering prayers for children, while children like me dared each other to pluck its leaves at dusk, convinced ghosts would leap out and follow us home. Exams, health, luck, curses—the tree carried responsibility for all. It was as if our entire village outsourced its uncertainties to that one restless canopy. What fascinated me was this duality. How could a single tree be both feared and adored? Over time, I began to see that they weren't opposites at all. The peepal embodied life and death, continuity and impermanence. Its roots clutched ashes from the cremation ground, yet its branches sprouted fresh leaves each dawn. The tree drank sorrow from the soil and exhaled oxygen more generously than any other tree.

Granny once said, 'Gods and ghosts are not enemies. They are both forms of belief. The peepal hosts them because it has space for both.' She was right in more ways than she knew. Ecology calls it *biodiversity*—the ability of a system to hold many lives, even conflicting ones, in balance. The peepal, with its hollow trunks, aerial roots, figs and nesting hollows, is nature's lesson on coexistence.

Granny's words echo even now: '*Ghosts are born when people believe too much or too little.*' I think that applies to nature as well. When we fear it excessively, we destroy it to feel safe. When we neglect it completely, it withers in silence. Between those extremes lies care—neither worship nor apathy, but attention.

Chapter 24

The Spreading Roots

The peepal tree also fascinated me for what lies beneath it: its roots. If you wish to understand the genius of survival, you must look underground. Beneath every peepal stands a world unseen, an architecture of patience and intelligence, a labyrinth of roots that listen to the earth.

The peepal does not depend on surface abundance; it creates its own. Its root system begins modestly, a single taproot descending like a pilgrim seeking faith in the dark. But very soon, it changes character. It sends out lateral roots in every direction, sometimes spreading three or four times wider than its canopy. They move through soil like thought through silence: touching, testing, sensing. They find what they need: water, minerals, space, and when they don't, they make do. These roots can survive drought and flood alike. They can grow through cracks in walls, hold on to rock, split open pavement and turn the hardest soil into humus over time.

I have seen peepal roots embrace entire ruins, temple stones, old wells and forgotten shrines, wrapping them in slow affection until the roots became one with the architecture. That is the peepal's way—it does not conquer; it coexists. It does not destroy the wall; it becomes the wall's memory. This is not aggression; it is evolution's quiet poetry.

Scientists describe the peepal's roots as fibrous, extensive and extraordinarily adaptive. The tree can thrive in sandy loam, black cotton soil, laterite or even gravel. It prefers slightly alkaline conditions but will tolerate poor, compacted or salty soils where most species fail. The secret lies in its *root plasticity*—the ability to remodel itself depending on conditions.

When the soil is rich and moist, it sends fine hair-like feeders to absorb nutrients; when it is dry or hard, it grows thicker, woodier roots that can reach deeper layers. It can take in oxygen, especially in waterlogged conditions, allowing the peepal to breathe where others would suffocate. These roots also form silent partnerships with microscopic fungi known as *mycorrhizae*. Together, they form an underground alliance that improves water absorption and nutrient exchange. The fungi receive sugars from the tree; the tree receives minerals from the fungi—a perfect ecological handshake. In this way, the peepal's roots are not solitary explorers; they are part of an underground community, alive with communication and exchange.

What fascinates botanists is the peepal's ability to combine three types of rooting systems in one life: taproot, lateral spread and aerial roots. The aerial roots that descend from its branches are extensions of its will to multiply. When they touch the ground, they thicken into pillars and become trunks themselves. Over decades, one peepal becomes many, forming a grove that appears like a family but is actually a single organism. Beneath the soil, these roots remain connected, sharing water, sugars and signals. In this way, the peepal creates a living cathedral of wood and wisdom, one being with many bodies.

Root spread in a mature peepal can reach an astonishing scale—20 to 30 metres in radius, and in old specimens, even more. Depth depends on soil type and water availability, often exceeding 10 metres in alluvial regions. But the tree's true power lies in its horizontal network, not its depth. These lateral roots act as natural anchors against wind and flood, holding entire landscapes together. Along highways and riverbanks, I have watched peepal roots bind loose earth and prevent erosion. In drought years, they preserve moisture beneath their own shade, creating microclimates that help smaller plants survive. Even the chemistry of the soil changes under a peepal. The constant fall of leaves adds organic matter, enriching the topsoil with carbon and nitrogen. The microbial life around its roots increases manifold. Over time, the soil beneath becomes darker, softer, more alive. In the language of ecology, it is a *climax species*, a tree that stabilizes and matures the ecosystem.

I have often thought that if the roots of the peepal were visible, people would worship them more than its leaves. They show us what true faith looks like: anchored, unseen and endlessly giving. The tree feeds from the dark to serve the light. It does not compete for survival; it collaborates.

Wherever I travelled, through plains, plateaus or temple courtyards, I found that the peepal's roots were at work quietly, reclaiming lost ground. In places where the soil had turned barren from neglect, they entered like healers, loosening compacted layers, inviting air and water back in. In old graveyards and abandoned fields, they were often the first to return, their roots nursing the soil back to life. They carry within them an ancient resilience that has adapted to India's monsoons, droughts and human forgetfulness.

Scientists say roots store isotopic evidence of rainfall and soil nutrients over centuries. In that sense, every peepal is also a historian. Its roots remember the chemistry of monsoons long gone, the minerals carried by vanished rivers, the footsteps of civilizations buried in dust. They are silent record-keepers of climate, a living archive of change and endurance. To study the roots of a peepal is to study patience itself. They do not hurry. They do not force. They persist. Over time, they find their way into everything: soil, stone, memory and myth. They teach us that strength need not be loud, that growth is not always upward and that wisdom often grows beneath the surface.

The peepal's roots are the Earth's handwriting, scripted in silence, fluent in survival. And perhaps that is why wherever I stand under its shade, I feel connected not just to the soil below, but to every invisible thread of life that runs through it. The roots beneath my feet are not only holding the tree, but they are holding the world.

Chapter 25

The Trembling Leaf

There is a particular moment, just before dawn, when the peepal tree seems to awaken before the rest of the world. Its leaves, heart-shaped, translucent and delicate, begin to shimmer as if the air around them were alive. Long before the first crow announces the morning, the tree whispers, not in sound, but in motion. Every leaf, an aerodynamic miracle, trembles, fluttering constantly as it catches invisible threads of air, creating a dance that has bewildered poets and biologists alike.

The peepal's leaves are simple, alternate, cordate and heart-shaped, tapering into a long, thread-like tip known in botany as the *drip tip*. This elongated apex is not decorative; it is functional. It allows rainwater to trickle off efficiently, preventing fungal growth and water stagnation on the leaf surface. Nature designs beauty with purpose. The petiole, the slender stalk attaching the leaf to the stem, is remarkably long, often 5 to 10 centimetres, and flexible. This is the secret behind the leaf's famous trembling. The petiole's flattened, flexible structure allows the leaf to pivot at the slightest breeze. It is a biological hinge, a marvel of nature's engineering. This gentle tremor maximizes air exchange, prevents fungal growth and allows sunlight to dance across every surface. While most trees rustle, the peepal shimmers, a silver-green flutter, constant even in still air. Botanists say it is due to the petiole's length and elasticity, but mystics have always known it as the tree's way of breathing.

If one looks closely, the leaf of the peepal is a story of veins and nerves. The primary vein runs straight like a spine from base to tip, with secondary veins fanning out symmetrically—a perfect geometry of life. In sunlight, these veins glow golden, like rivers on a green map. As a child, I would hold the fallen leaves against the morning sun and trace their patterns with my finger. Granny would say, 'Every leaf is a book, Baba. You just have to learn its script.'

Even after drying, peepal leaves remain graceful. When soaked and rubbed gently, the green pulp peels away, leaving behind a skeleton, the intricate lace of veins, delicate as lacework. Artists have used these skeletal leaves for centuries, as natural canvases for miniature paintings, prayers and poems. I once tried writing on one with a fountain pen; the ink bled, but the leaf didn't tear. It was like writing on time itself.

Scientifically, peepal leaves are powerhouses of photosynthesis. Their large surface area and thin lamina allow efficient gas exchange. Under a microscope, the stomata of peepal leaves resemble tiny mouths that open and close rhythmically. I've always found it fascinating that something so silent, so still, could be so busy, so full of purpose. When you sit beneath a peepal, you are sitting inside a vast biological engine converting sunlight into life.

In Hindu mythology, the trembling of peepal leaves is not mere movement; it is a divine communication. It is said that the Gods dwell in its rustle and that the air around it is filled with subtle vibrations of consciousness. The Bhagavad Gita speaks of the peepal as the cosmic tree—the Ashvattha—whose roots are in heaven and branches on earth, a symbol of eternal life. Buddha attained enlightenment under the peepal tree in Bodh Gaya, the *Bodhi vriksha*. The same trembling leaves that catch the wind also catch light from other worlds.

Granny used to tell me that when the peepal trembles at midnight, spirits pass by. When it trembles at dawn, Gods descend to bless the earth. I would sit awake, watching the leaves in moonlight, and wonder which of the two was true. Over time, I realized both were, for the divine and the ghostly are the same forces of nature, misunderstood in different lights.

The peepal leaf has its own scent: faintly resinous, earthy, with a hint of green tea and rain. When crushed, it releases latex, a white sap, which protects wounds and seals cuts. In Ayurveda, powdered peepal leaves are used for asthma, and their decoction for heart ailments. The leaves are nature's first aid, humble yet potent.

Interestingly, the peepal's leaves are acutely sensitive to pollution, functioning as living sensors of the air we breathe. The constant fluttering, even in still air, enhances the interception of airborne particulate matter. As vehicles pass and exhaust rises, microscopic particles of soot ($PM_{2.5}$ and PM_{10}), along with heavy metals like lead, cadmium and nickel, settle onto the leaf surface.

Scientifically, this process is known as dry deposition. The slightly rough, waxy cuticle of peepal leaves traps dust particles, while the stomata absorb gaseous pollutants such as sulphur dioxide, nitrogen oxides and ozone. Over time, these pollutants accumulate, darkening the leaf surface. What we casually perceive as 'dirty leaves' is, in fact, evidence of silent environmental service.

In cities like Delhi and Lucknow, especially at traffic intersections, peepal trees often appear visibly darker than those growing in rural landscapes. Studies have shown that trees growing along high-traffic corridors accumulate significantly higher concentrations of heavy metals in their leaves than those in cleaner environments. The peepal does not reject this burden; it adapts. Pollutants are either immobilized within leaf tissues or later shed when the leaf falls, effectively removing contaminants from the immediate air column.

The trembling of peepal leaves, so often romanticized in folklore, has a physiological role as well. This constant motion prevents the formation of stagnant air layers around the leaf surface, improving gas exchange and increasing the tree's efficiency in absorbing pollutants. It is nature's version of active filtration, achieved without electricity, maintenance or complaint. Yet this service comes at a cost. Polluted leaves show altered chlorophyll content, reduced photosynthetic efficiency and signs of oxidative stress. And still, the tree persists. Season after season, it continues to clean the air, shade the streets, cool the surroundings through transpiration and offer oxygen in return for neglect.

There is something profoundly moral in this exchange. While urban life rushes past—horns blaring, engines burning—the peepal stands rooted, absorbing what we release, holding onto our excesses and asking for nothing. Its leaves darken so our lungs may breathe a little easier. Its silence is not passivity, but endurance. It is almost as though the peepal forgives our sins quietly, one leaf at a time.

In my life of planting, I've handled millions of peepal leaves, young ones, glossy and soft like newborn palms; old ones, brittle, veined like aged hands. I've seen them fall, cover the ground like pages from an ancient manuscript. Sometimes, I think of all the trees I've planted, and imagine their leaves fluttering together, an invisible orchestra of trembling hearts. And so, whenever I sit under a peepal tree and see its leaves shimmer in the faintest breeze, I am reminded of what my Granny once said: 'Even when the world stands still, the peepal listens.'

Chapter 26

Chlorophyll Is God

It was in the early 1980s that I began following the discourses and taped talks of a new, controversial orator and philosopher, Acharya Rajneesh. His voice was hypnotic, his logic razor-sharp. Every sentence seemed to peel away the pretensions of religion and society. I devoured every article that appeared in newspapers and magazines, in Hindi and English alike. He was building an ashram in Pune then, and his words felt like a storm cleansing the dusty corridors of Indian thought.

Acharya Rajneesh, later known to the world as Osho, was a professor of philosophy in Jabalpur, Madhya Pradesh. He left academia to create a space where meditation and rebellion could coexist. I admired his courage to challenge the system. He demolished hollow rituals, exposed political hypocrisy and spoke about freedom, nature and consciousness in a language that was alive. To me, he was a hero—Batman, Phantom, Spiderman and Superman all rolled into one. My relatives, of course, declared I was hypnotized. They were probably right.

At that time, I was studying at Allahabad University, and I had developed a fascination for non-fiction, philosophy, ancient history, sociology and anthropology. I read them like others read comic books. I would spend hours in second-hand bookshops, flipping through tattered pages that smelled of dust and wisdom. Then came Osho's discourses on ecology and the living earth. He said something that still echoes within me: '*Nature is God in action. God is nature, and nature is God.*' It struck me like lightning. He would say, '*If you cannot love a tree, you cannot love existence.*' Those simple sentences rearranged my priorities forever.

Chlorophyll Is God

'*Patta hi Parmatma hai*'—the leaf is God—said Osho. Though granny had said it before, he gave it a new depth, a universality that pierced through mysticism into science. When he said, '*Chlorophyll is God,*' he was not being poetic. He was pointing to the core of existence, to the green pulse that keeps the planet alive.

Chlorophyll is the divine bridge between sunlight and life. It is the invisible priest conducting the marriage of light and matter. Through it, the sun's energy is converted into sugars, into food, into forests and into us. Photosynthesis, the quiet miracle that happens when chlorophyll dances with sunlight, creates the oxygen we breathe. Every breath we take is a continuation of that divine transaction. We inhale what the leaves exhale; we live because they live. The sacred cycle is not a metaphor; it is physics, chemistry and spirituality intertwined.

The green colour of leaves is not just a pigment; it is the visible form of divinity at work. Every tree, every blade of grass, every algae bloom in the ocean is a temple where this worship of light takes place. When you walk under a canopy of peepal, neem or jamun, you are walking inside a cathedral of oxygen. The leaves are chanting in silence. Their hymn is the exhalation that fills our lungs.

Photosynthesis began long before language, long before Gods had names. It began when the first green cells learned to drink sunlight and release life into the air. That was the original prayer, the first act of creation on Earth.

When Osho said 'Patta hi Parmatma hai', he meant that the divine is not elsewhere; it is in the chlorophyll molecule, in the tender leaf trembling in sunlight, in the greenness that sustains breath. Every act of planting, every shade we create, every garden we nurture is therefore a spiritual act. The gardener becomes a priest, the sapling becomes scripture and photosynthesis becomes prayer. So, when I look at a leaf now, I bow, not to mythology, but to biology; not to an idol, but to life itself. Chlorophyll is not just God; it *is* the reason God exists in us.

When I finally reached the Pune ashram, I felt I had entered a different planet, one overflowing with dance, meditation, laughter, silence and greenery. What drew me most were the trees: ancient banyans standing like guardians of wisdom, and groves of bamboo whispering like monks in the wind. The ashram was an oasis, with ponds, waterfalls, gardens, rockeries and birds everywhere. The entire campus was alive, pulsating with the rhythm of nature. Even the air seemed to hum with energy. My lifelong love for plants had finally found its temple.

Osho had a deep affection for bamboo, calling it 'the enlightened plant'. Flexible, hollow, upright and humble, the perfect metaphor for the human being who has emptied himself to allow the divine to flow through. I remember his words vividly: '*Be like the bamboo—open, receptive, flexible. When the wind passes through, it sings.*'

Osho spoke often of ecology as a spiritual balance, not environmental activism. He said, '*The trees are your lungs, the rivers your blood, the earth your body. To hurt them is to hurt yourself.* I realized then that planting a sapling was not a social act; it was a spiritual offering. Meditation and plantation became one movement in my life. And so, another sannyasi was born. I was given the name Swami Prem Parivartan. Swami became my first name, Prem (meaning 'love') my middle, and Parivartan ('transformation') my last. I wore maroon robes, a mala around my neck and carried the easy smile of someone who had just been reborn.

I was, of course, born as Azad Jain on 26 January, a day of freedom. My family has always been fond of strong, rebellious names—Kranti, Swatantra, Yudhanti, so 'Azad' was a natural choice. My father's Jain roots trace back to Rajasthan, and my birth certificate still carries that lineage.

Over the years, 'Swami' naturally replaced 'Azad'. My documents followed suit. One by one, my bank account, ration card, driving license, Aadhaar and voter ID all quietly transformed. It was as if the universe wanted me to let go of the old and embrace the new.

By 2003, I made it official through the *Gazette of India*. The name Swami Prem Parivartan became legal and complete. My nickname, Peepal Baba, had already gained popularity, and people often got confused between the names. Once the papers were done, the confusion ended, and the banks were happier too.

Today, people call me by many names: Swami Sir, Swamiji, Mr Prem, Mr Parivartan or simply Swami Uncle. I let them choose. After all, true freedom lies in giving others the freedom to call you what they wish.

Looking back, I realize that the ashram was never just in Pune; it was everywhere I carried my saplings. What began as meditation under banyans slowly became fieldwork under the open sky. The robes changed colour, the cities changed shape, but the silence within remained the same. The *sanyas* I took in the ashram found its real meaning on the roadside; in dust, sweat, laughter and the company of friends who believed that change, like love, must grow leaf by leaf. In that sense, Osho never left me. He simply turned into the trees.

Chapter 27

The Economy Beneath the Canopy

There are mornings when I ride out without a destination, just following the roads that curve toward shade. That's where the stories are. Shade has always been India's oldest currency, traded daily between trees and people without a receipt or signature. Beneath every broad canopy, life unfolds like a marketplace that doesn't need walls.

I have seen the country from under those trees. From Delhi's dusty bylanes to Bundelkhand's parched squares, from the sleepy edges of Bihar's stations to the glowing chowks of Rajasthan, the peepal tree stands like an ancient accountant, quietly keeping the ledgers of human survival. Beneath it, economies hum that no ministry has ever counted.

At dawn, the first arrivals are the chaiwallahs. They come pushing their wooden carts, rattling tin kettles like temple bells. The peepal is already awake. They tie a rope between two roots, hang an old bulb from it and the day begins. The scent of boiling tea rises through the leaves; customers gather; gossip and trade flow like two sides of the same river. By mid-morning, the barber arrives, his mirror catching sunlight in small, playful flashes. Next to him sits the cobbler, hammering soles into new life. They never fight for space. The canopy gives enough for everyone. I once asked a chai vendor near Mathura why he always chose a peepal tree. He looked up and said, '*Yeh ped chhat bhi hai, saathi bhi* (This tree is both a roof and a companion).' That's the economics of emotion, a lesson no textbook can teach.

The peepal tree is not just a species, it is infrastructure. It replaces tin roofs, shop rentals and electric fans. It cools the air, reduces noise and gathers people into a micro-community. Every peepal becomes a village square, a parliament, a canteen and a co-working space, all in one. Across India, panchayats still gather beneath peepal trees, not in concrete offices. Disputes are settled not with law books but with common sense. The air itself seems cooler when people talk under a peepal. The voices mingle, egos dissolve and truth often arrives like a breeze, unexpected yet cleansing. In the dry interiors of Madhya Pradesh, I've seen weekly *haat*s form around peepal and banyan clusters. Women spread vegetables on gunny bags, weighing them with rusted iron stones. Children nap in the branches while elders trade grain for utensils. The tree keeps the sun in check, like a silent regulator of fairness. If the same market moved to an open ground, half the sellers would faint by noon.

In the cities, too, the pattern repeats. On Delhi's Ring Road, near every traffic light, you will find small worlds under peepal trees: flower sellers, fruit vendors, beggars and mystics, all coexisting like different species in a thriving ecosystem. Even in Lucknow, I once saw a mechanic set up his entire workshop under a peepal—jack, tools, and tyres spread neatly around the roots. When I asked him if he was worried about rain, he smiled: '*Yeh ped mujhe dhoka nahi deta. Dukaan se zyada yakeen is par hai* (This tree never deceives me. I trust it more than a shop).'

A Hidden Employment Network

Somewhere in those years of wandering, I began counting, not for statistics, but out of curiosity. I noticed that nearly every large peepal tree sheltered at least ten families. Vendors, priests, fruit sellers, tea makers, palm readers, key-makers, barbers and mechanics, all orbiting one canopy. If each family supported four dependents, then one tree sustained forty people. Multiply that across millions of peepal trees across the country, and you begin to see the unseen economy: a living, breathing employment scheme run on photosynthesis. Economists may call it informal work; I call it *organic livelihood*. No rent, no electricity, no paperwork; just trust, shade and human rhythm. The peepal is a socialist institution without a signboard. It never discriminates by caste or class. A paan seller and a holy man can share the same shade without argument. Beneath its branches, the nation's most egalitarian workspace unfolds quietly every day.

Sometimes, the peepal stands near a temple, and commerce blends with prayer. I have seen garland-makers stringing marigolds while children chase cows around the roots. The priest conducts rituals; the barber sharpens his razor nearby; a beggar sings bhajans in a broken voice and someone always brings a *matka* of cold water for all. Nobody planned this; it grew naturally, like the tree itself.

Economics of Shade

If you were to assign a value to that shade, it would astonish you. One canopy saves a vendor from paying daily stall rent, roughly Rs 200 a day. That's Rs 70,000 a year in avoided cost. Multiply that by millions of such trees, and the national contribution reaches the scale of formal industries. Yet, no economist counts it. The peepal's GDP is invisible because it is compassionate. It does not ask for rent, license or electricity bills. It simply gives, and in doing so, sustains a parallel economy built on reciprocity rather than competition.

Urban planners, however, often see these clusters as encroachments. They trim roots, erect fences and drive away the vendors in the name of beautification. What they remove is not clutter, but community. When a tree is cut, it is not just wood that is lost; it is an entire livelihood network, an ancient economy of coexistence.

On a highway near Fatehpur, I once stopped by a peepal that had grown through the cracks of an abandoned well. Around it, four families lived, selling cold drinks, fixing punctures, frying pakoras and running a barber's chair. They said they never paid for electricity; the tree was their fan. The roots had become benches, the branches their roof and the leaves their lullaby at night. One of them told me proudly, '*Baba, hum log ped ke kiraye ke mazdoor hain* (We are tenants of the tree).'

In another town, a boy studying under a peepal's shadow used its trunk as his desk. When I asked his mother what she did there, she said she sold incense and flowers to passersby. Between her trade and his studies stood one truth: the tree was their school, shop and shelter, all in one.

I have also seen these trees by railway stations, guarding time itself. At smaller stations in Uttar Pradesh and Bihar, the peepal's tower is usually beside the platforms, shading chai counters and bookstalls. The trains come and go, but the tree never departs. It witnesses generations of vendors, policemen and travellers, holding all their lives within its breath.

If you travel long enough through India, you realize that the country runs not only on power grids and highways, but on these green parasols scattered through towns and villages. They are the cooling towers of human hope. They provide more employment than many factories, more peace than many offices. In the evenings, when the vendors pack up, the same space turns into a resting ground. Stray dogs curl near the roots, old men gossip, someone lights a diya and children chase shadows. The economy gives way to ecology, and ecology to silence. The peepal doesn't sleep; it listens.

I wonder what would happen if economists spent one afternoon under a peepal tree—not in their offices, but out there, amid the smell of dust, tea and sweat. They might see what I have seen: that the nation's oldest enterprise still runs on sunlight, shade and faith. No balance sheet can capture it, but every chaiwallah knows its value.

Granny used to say, '*Beta, pedon ke neeche log sirf thand nahi lete, rozi bhi lete hain* (Under the trees, people find not only shade, but livelihood too).' That single sentence is the summary of this economy beneath the canopy, and perhaps of my life's work. Trees are not just our lungs; they are our livelihoods. And if you sit long enough beneath a peepal canopy, you will hear it whispering the oldest economic truth of India: *that prosperity, like shade, multiplies only when shared.*

Chapter 28

Grasses

I have always taken help from the grasses. Whenever we received a new plantation site, my first instruction to the team was simple: *Do not touch the grasses*. They were there long before us, long before the saplings we planned to plant and long before any of our human dreams took root in that soil. Grasses are the ancestors of that land. They are the first caretakers, the quiet healers who never ask for attention. If there was any native vegetation already growing, we let it be. We never wished to disturb the spirits that lived in those blades of green. Sometimes, when the growth was too tall, we shaved the grass lightly, just enough to make space for our saplings to breathe, but we never uprooted them. Pulling them out by their roots always felt like an act of violence against the very soil that was ready to welcome our trees.

Grasses are the doctors of the earth. Their fibrous roots hold the topsoil in place when the rains lash and the winds rage. They stitch the land together when erosion tears it apart. Many of them fix nitrogen and prepare the soil for the trees to come. They invite fungi and microorganisms to build invisible cities underground—vast networks that carry messages, nutrients and memories from one root to another.

Whenever I saw a site covered with wild grasses, I knew that the soil beneath was still alive. The presence of grasses meant hope. It meant that nature had not yet given up on that patch of land. In barren or degraded areas, we always used grasses as the first line of recovery. They are the green nurses that prepare the ground for the larger species to take their place.

I have learned from the grasses the meaning of fortitude and humility. They bend but never break, they grow where others cannot; they survive trampling, grazing and fire, and yet, they return every monsoon with renewed strength. Their simplicity hides deep wisdom: to live close to the ground, to rise again after every cut and to give life to others silently.

India's grasslands are among the most underappreciated ecosystems, and yet, they are the hidden foundation of our forests, rivers and agriculture. They are nature's first responders, the pioneers that reclaim disturbed land and nurse it back to fertility. I have immense respect for the grasses. They have been my oldest companions in this long journey of greening the land—silent, steadfast and sacred.

Grasses like vetiver, known locally as Khus, are nothing short of a miracle. Its roots go as deep as 3 to 4 metres into the soil, binding it like invisible ropes. We often planted vetiver along slopes and bunds to prevent soil erosion, runoff and even filter pollutants from wastewater. When I stood on a reclaimed hillside years later, I would always find vetiver standing guard, unshaken by wind or water. Its fragrance, hidden underground, felt like the perfume of resilience itself.

Doob, the most common of all Indian grasses, is both sacred and scientific. Used in temple rituals and homa ceremonies, it also has extraordinary regenerative powers. Once it takes root, it spreads horizontally, binding loose soil and reducing dust. At many barren sites, our first green came not from tree leaves but from a carpet of Doob that spread naturally after the first rain. Doob grass is humility made visible, small, silent and persistent.

Napier also stabilizes embankments, preventing them from crumbling into dust. Tall and generous, it not only protects but also nurtures, feeding cattle across the country. In community projects, we used Napier as a boundary species. It grew fast, gave fodder and prevented animals from straying into young plantation zones. It became a friend to both farmer and forest.

Another grass used in all sacred rituals is Kusha. It is mentioned in the *Rig Veda*, but beyond its spiritual aura, it is an ecological marvel. It thrives in poor, sandy soils and its roots bind dry, arid lands where almost nothing else grows. Wherever we found Kusha grass growing naturally, we treated that land with reverence. It meant that the earth still remembered her strength.

In later years, we began using lemongrass and citronella on degraded urban sites. Their fragrance kept insects away, their roots stabilized the ground and their leaves could be harvested without harming the plant. In time, they invited bees and butterflies.

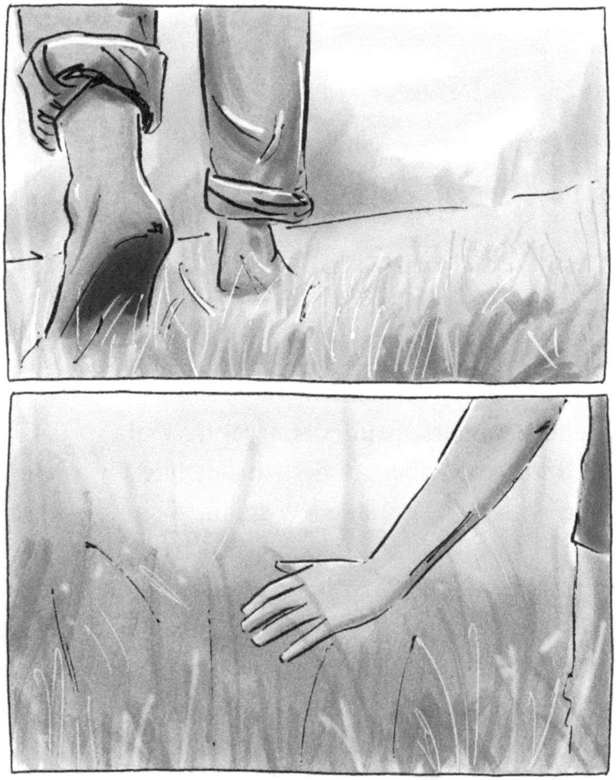

When you walk among grasses, you are walking through a living philosophy. They teach you everything a planter must know: to stay grounded, to work quietly, to heal before claiming space and to never give up even after being cut down. The grass does not shout like a tree, nor does it demand worship like a flower. It just grows—in cracks, on pavements, in dust, in sunlight, in shadow. It grows where it can, not where it is praised. Whenever I kneel to plant a sapling, I touch the grass around it and whisper, 'Take care of this one.' They always do.

Chapter 29

Of Volunteers, Gurus and Teachers

Children, strangers, students, sceptics—they all slowly blurred into one tribe. At first, it felt like I was the lone madman with a red cycle, a few saplings dangling in a bag and a wild idea that roads deserved shade. But soon the madness spread, like a rumour too delicious to ignore. Nobody announced it, nobody signed forms, nobody appointed leaders. Yet, without notice, we became something larger. We became a ground force. Not the kind with uniforms and medals, but with muddy palms and sweaty brows. Our anthem was the rustle of new leaves, our flag the crooked silhouettes of saplings swaying against the evening sky. If you squinted hard enough, you could see us—children stomping soil, a vegetable vendor squeezing out leftover water onto roots, a rickshaw puller clanking his tin, students cheering louder than they dug. We weren't an NGO, we weren't a society, we weren't even sure what we were. But the road knew. The saplings knew. And secretly, I think, the sceptics knew too.

There was one woman who sold vegetables by the roadside. She had no money to spare, and hardly any time between chopping and bargaining, but every evening, she poured her leftover water from washing spinach and cauliflower onto a sapling nearby. Once she said to me, 'I cannot plant, but I can give what I have.' And that small act, repeated daily, kept the sapling alive longer than speeches ever could.

There was a rickshaw puller who tied an old tin to his cycle, filling it from roadside taps and watering trees along his route. When I asked him why, he shrugged and said, 'When these trees give shade, my sons will rest here. That is enough payment.'

Children became self-appointed tree guards, chasing goats away with shrieks and sticks, puffing their chests with pride as if they had saved kingdoms. One boy took ownership of a jamun sapling and announced to all that no goat in the colony would dare nibble it as long as he was alive. The goats, of course, had other plans, but the boy's loyalty was unquestionable.

Planting became a celebration. Birthdays were no longer about balloons or cakes but about digging holes. 'Plant a tree instead of cutting a cake,' I would suggest, and to my delight, people listened. Exam results, anniversaries and even quarrels were sometimes settled with a sapling pressed into the soil.

The roadside became a temple of sorts, not with incense or fire but with soil and water as the rituals. Some tied sacred threads around trunks, whispering wishes into leaves. Some brought flowers to decorate saplings as if they were idols. The trees became ecological and spiritual companions at once. The Gods must have been watching, amused, because our rituals grew without priests, without scriptures, yet with an honesty hard to find in temples.

Saplings, for me, were never just plants. They were children—fragile, dependent, demanding care, unable to defend themselves. They cried in silence when thirsty, drooped like exhausted infants in summer heat, shivered in winds they could not resist and withered when neglected. And just like children, they rewarded you generously if you loved them sincerely. Few joys compare to the sight of a sapling pushing out its first new leaf. After weeks of carrying water in dented cans, guarding against goats with laughably small sticks and pleading with neighbours not to dump waste on fragile roots, one morning you arrive to see a tender green shoot. Lighter than the rest, trembling with courage, pointing skyward. It feels like hearing a child's first word. My smile would stretch so wide that passersby must have thought I had won the lottery.

Sometimes I even spoke aloud to the sapling: 'Well done! You survived.' People might laugh at such a conversation, but I didn't care. Talking to saplings was as natural to me as scolding a stubborn cousin or coaxing a sulking friend. Granny had taught me long ago that plants listen if you whisper. And if Granny said so, it was the law.

But with joy came grief. Many saplings did not make it. Goats chewed them, carts rolled over them, careless feet snapped them, relentless summers withered them. Each loss was a stab. Once, I had nurtured a neem with special devotion, guarding it with stones, whispering encouragement daily. One afternoon, I found it snapped in half by a cart's wheel. I stood staring at the broken stem, as if I had lost a sibling. People say grief over a plant is childish. But those who have planted and cared know that grief is real. It is as sharp as any other loss. To ease my pain, I began naming them. A guava became Hope. A peepal became Strength. A neem became Patience. Visiting them by name made my care personal. When one died, it wasn't 'a plant lost'; it was Hope gone or Patience gone. My roadside family shrank and grew in unpredictable rhythms, like life itself.

The more I cared, the more I guarded. I built small enclosures from sticks and stones, ringed saplings with thorny branches to keep goats at bay. Sometimes I even stood nearby, arms waving madly at approaching cattle, looking like a scarecrow who had misplaced his farm. At night, I worried: Had the heat been too much; had the rains drowned the roots; had some reckless hand uprooted my fragile green wards? That kind of protectiveness was nothing less than parental. Responsibility, I learned, is not a dictionary word; it is the weight in your chest when you know something fragile depends on you.

But what gave me real strength was that I was never alone. The ground force had adopted the saplings too. They were there to multiply the joy when our saplings survived and to share the burden of loss when they didn't. Everyone had their favourite. The vegetable vendor's daily offering, the rickshaw puller's shade-dream, the children's jamun guard, all were acts of parenthood. Together, we became a family of caretakers, bound not by blood but by roots. In our little republic, trees were the citizens, and we were the humble servants.

Volunteering began as help and became a homecoming. People came with curiosity, stayed with conviction and left behind something more enduring than footprints—roots. I never built an organization; I built a circle. There were no uniforms, no roll calls, no banners. Only the rhythm of digging, planting, watering, waiting. One person came to escape loneliness, another to fill a Sunday, a third because someone else had dragged them along. But somewhere between holding the sapling and touching the soil, something shifted. Volunteering stopped being a weekend habit and became a way of breathing. From one man with a cycle, it became dozens, then hundreds, each with their own style of madness. Children with muddy hands, women with steel lotas, men with spades, policemen who waved me through, rickshaw drivers who ferried extra water, all became part of the story. It was less an organization and more an epidemic. One person planting trees can be dismissed as eccentric. Ten planting trees becomes a nuisance. A hundred people planting trees suddenly begins to look like a movement. But movements are never tidy. Some volunteers arrived late and left early. Some planted one tree and declared themselves environmental revolutionaries. Some planted with such force that they nearly uprooted the sapling immediately. Yet, I cherished every hand, every attempt, every crooked pit, every overwatered root. Because trees do not demand perfection, they demand persistence. And people, with all their clumsiness, supplied just enough of it to keep the green dream alive.

I think of Ramesh, a college boy with restless limbs and loud laughter. He treated our planting drives as adventure sports. But over the months, I saw his swagger turn into stillness. He began guarding the saplings like a priest guards a shrine. He scolded goat herders, chased away litterers and once told me, half-angry, half-proud, 'Sir, this is not your tree anymore, it's mine too.' That sentence remains one of the finest compliments I have ever received. For a tree grows roots in human hearts long before it anchors in the soil.

Shalini, soft-spoken and stubborn, adopted a gulmohar on a dusty roadside. Every week, she brought a bottle of water and a pocket of conversation for it. 'It's my friend,' she said. 'When I'm sad, I sit under it. When I'm happy, I tell it first.' She had begun to speak the language of silence, the dialect of the Earth.

The faces kept changing. Students, retired officers, housewives, taxi drivers, monks, corporate interns. Some came for a day, some stayed for years. They all left fingerprints on the bark of time. Our work had no pay, but it had plenty of reward—the kind that grows quietly in the chest when you see a leaf where there was once only dust. Slowly, volunteering became our way of praying with our hands.

There were other teachers too: fire, drought and failure. Once, a small fire reduced a line of neem saplings to blackened sticks. We thought the dream had ended. But the next monsoon, green shoots emerged again, alive and laughing. The Earth rehearses resurrection every season. She never gives up, so why should we? Digging a pit became our meditation, removing stones of ego, weeds of impatience and burying doubt like dry leaves. The spade was our scripture; sweat our prayer. Someone asked, 'Sir, when will this become a forest?' I answered, 'The forest begins inside you. The day you care for one tree like your own child, it's already here.'

Work grew, as work does when done with love. We were no longer volunteers; we were a community of listeners. The Earth spoke in the crack of a seed, the flutter of a leaf and in the silence of roots. I told my volunteers: The planet doesn't need saviours, it needs students. Every morning, the world begins again, and the syllabus is written in soil. Those who attend her class learn patience, humility and laughter that smells faintly of compost.

We discovered that every act of care multiplies. One broom, one bucket, one seed—each carries a universe inside. 'First clean, then green,' I would tell them. The broom and the sapling are partners. You cannot plant in filth. Children once saw me sweeping a site before planting and whispered, '*Yeh uncle jhadu kyun maar raha hai?*' When I planted the saplings later, they clapped. The broom had justified itself.

Some lessons came from people we never expected. In Kotada Village, an old farmer pointed at the soil, then the sky, then his palms. That one gesture explained agriculture, faith and patience all at once. In another place, a boy wrote in his diary, 'Thirsty soil.' Two words, but they held the ache of a whole summer. Years later, reading that line, I could still feel the cracked land waiting for rain. I keep telling my team: Write it down. Don't trust memory, it fades faster than footprints in dust. Field notes are treasures. Thoughts need diaries, the way seeds need soil.

Sometimes, after the day's work, we sat quietly under a peepal. The volunteers joked that the leaves gossip like aunties. I said, 'Yes, and they tell the truth.' The peepal, to me, has always been a living classroom. Sitting under it, the world slows down.

Across forty years, I have met thousands of volunteers, children who grew up to become teachers, doctors, soldiers and artists. Some still send photos of trees they planted decades ago. Their messages begin the same way: 'Sir, it's still alive.' That sentence is worth more than any award. It means that the work took root in soil and in soul.

We travelled through twenty-one states and over two hundred districts, collecting not just saplings but stories. I've learned that the soil of Odisha smells different from that of Rajasthan, that children in Kerala can make a song out of planting and that a farmer in Uttar Pradesh can deliver more wisdom in one shrug than a thousand seminars.

These journeys were not tourism; they were pilgrimages of sweat, mud and patience. When I say shade came to six lakh acres of land, I don't say it like a statistic. I say it like a traveller who found a tree after cycling in the May heat.

The world calls it volunteering. I call it returning a debt. We owe the Earth more than gratitude; we owe her participation. To dig, to plant, to care, to fail, to try again. That's the real community work: to show up, again and again, like the monsoon, uncertain but full of promise. The volunteers who came and stayed are not just part of my story; they are my forest. Because in the end, movements don't grow from manifestos. They grow from muddy hands, shared laughter and faith that shade will come, if not for us, then for someone yet to walk this road.

Discovering My Gurus

It was in those years that I discovered my gurus, though at the time, I didn't call them that. Granny had been my first. Rising before dawn, whispering to Krishna with quiet devotion, she showed me that prayer is not a grand affair but a rhythm, like breath. When she scolded me for hoarding toys and ordered me to give them away, she taught me that simplicity is wealth. When she laughed at my fears and pressed a Hanuman idol into my hand, she showed me that fear is real, but faith is stronger. If a guru is someone who opens your eyes, then Granny was mine. Sometimes I wonder if it was those lessons in simplicity and strength that brought in volunteers who gave so much to the cause.

I remember sitting under a peepal tree with her after she had told ghost stories to the children. They had run off shrieking and giggling, but I stayed. She sat silently, eyes on the swaying branches. 'Why are you silent, Granny?' I asked. She smiled: 'Because silence has more to say than words.' At that moment, I was confused. How could silence speak? But as I sat beside her, I began to hear the rustle of leaves, the koel's call, the faint bark of a dog and somehow, they all blended into something fuller than noise. That was the first whisper of the inner guru, not from outside, not in books, but in the silence that appears when you stop chattering.

If Granny was my first guru, then Gandhiji was the second. He was everywhere in India's air: on currency, in textbooks, in the slogans painted on walls. I first heard about the Salt March from a teacher, who thundered, 'Imagine a man walking hundreds of kilometres just to make salt, against an empire!' I sat stunned. No soldiers, no weapons, just bare feet and faith.

Later, reading *My Experiments with Truth*, I found a man confessing mistakes with honesty instead of boasting of victories. That honesty was more powerful than polished speeches. Gandhi taught me that truth is a compass, not a chapter heading. Even in planting saplings, I had to face the truth: some lived, some died and pretending otherwise was useless. He also taught me simplicity. Possessions don't strengthen; they weaken. Joy does not come from gadgets; it comes from breathing clean air, seeing saplings grow and hearing birds return. When neighbours mocked me, when vandals uprooted plants, Gandhi's lesson of nonviolence steadied me. My revenge was persistence. My protest was replanting. Service, too, became worship. Bending to press soil, carrying water cans, persuading children to help, these became prayers more authentic than incense. The roadside was my ashram, the sapling my idol.

And then came Acharya Rajneesh, like a monsoon storm. If Gandhi was the whisper of restraint, Rajneesh was the roar of freedom. He mocked priests, laughed at seriousness, told stories that began as jokes and ended as revelations. His cassettes reached me in dusty covers, and when I pressed play, his voice poured out, sharp yet gentle, playful yet profound. He said meditation is not suppression but celebration. Dance, laugh, sing, cry, then fall into silence. I tried his dynamic meditations clumsily, feeling ridiculous, but afterwards the silence was deeper than any forced quiet. Suddenly, even planting saplings became meditation. Digging holes, pressing soil, watering roots—all had rhythm, breath, surrender.

Acharya Rajneesh taught us that the truth is not neat; it is a diamond with many facets. He urged us to unlearn, even to drop him when the time came. That was frightening but liberating. Certainty is comfortable; questioning it is terrifying. But his laughter carried me through. Spirituality no longer felt heavy; it felt playful, like Krishna stealing butter in Granny's tales.

Between Granny, Mahatma Gandhi and Acharya Rajneesh, I realized something simple: none of them was the guru. They were fingers pointing to the moon. The guru was inside—the voice that woke in silence under the peepal, that insisted on truth while carrying water, that laughed at contradictions while planting saplings. The roadside was my classroom, the saplings my classmates and my workforce was my congregation.

Looking back, I see a corridor of green where once there was only dust. I see children who laughed and then planted, strangers who mocked and then watered, neighbours who sneered and then protected. I see saplings that lived and others that died, each teaching me joy or grief. I see Granny's riddles flowering, Gandhi's truth rooting, Rajneesh's laughter dancing. I see a boy on a red cycle who thought he was planting trees but who was really being planted himself, into faith, into resilience and into freedom. And I can't help but smile.

Perhaps Granny was right: *ghosts do linger*. Because even today, as I walk down shaded roads, I feel them—the ghosts of children's laughter, the ghosts of Gandhi's sandals on salt, the ghosts of Rajneesh's chuckle and the ghosts of Granny whispering, '*Silence has more to say than words*.'

Chapter 30

Fieldwork and Funding

'Nature never did betray the heart that loved her.'

—*William Wordsworth*

Everywhere I go, people ask about the challenges, hurdles and impossibilities that must have stood in my way all these years. But when I look back, I find it hard to name them as hurdles. Every difficulty turned into a doorway, a new relationship, a milestone, an opportunity, a new location or the possibility of another forest.

I have never been shy of asking for help. Being expressive, talkative and transparent turned out to be an asset. People came forward, sometimes with money, sometimes with material, sometimes just with faith. Over the decades, those gestures became our invisible ecosystem of support.

How We Funded Ourselves

For more than forty years, we survived on microdonations. Small acts, large hearts. Food packets, sacks of rations and accommodation were often sponsored. Panchayats contributed fencing wire, manure, irrigation pipes and water tankers. Rich farmers lent us their tractors and trolleys. Some gave us crates or a lift on a scooter. Others gave us a few coins or an hour of help watering in the heat. Volunteers pooled petty cash to pay the labourers or buy saplings. Sometimes we had to fund every nail and nut ourselves. At other times, everything—from the shovel to the snack—was sponsored by someone who appeared from nowhere.

When you do the right thing, the right people appear. Good deeds attract good hearts. People are essentially giving, but they don't always know *where* or *whom* to help. What they seek is trust, and when they find it in consistent, verifiable work done with honesty, they open their doors.

A Community of Supporters

Our donors were not the wealthy elite but the everyday heroes of India: autorickshaw drivers, cab drivers, delivery boys, bank clerks, civil servants, teachers, factory workers, farmers, shopkeepers and fellow volunteers. Some contributed Rs 10, some a tiffin box, some their Sunday mornings. Many donated old cycles, scooters or motorcycles—our most practical and beloved transport for carrying saplings, manure and tools. Some brought chai and biscuits for our teams at the sites. Others stood silently, clapping or offering a word of encouragement. Every gesture mattered. The bucket never ran dry because everyone kept adding their drop.

Sometimes, donors gave us cows, not out of charity, but out of understanding. Cow dung has been our quiet treasure. When dried and treated scientifically, it restores degraded soil better than most chemical interventions. Farm waste, kitchen waste and garden waste from neighbourhoods were all transported to our sites. We made compost heaps, turned them lovingly and created the humus that birthed new forests.

For four decades, from 1977 to 2017, we lived purely on community goodwill and individual donations. It was only in 2017 that HCL Foundation extended its first institutional funding. Until then, it was faith and fieldwork that paid our bills.

But we were never alone in our work. The wind, rain, soil and the goodwill of strangers became our invisible partners. There were always people advising, instructing and correcting, not out of expertise but out of habit.

Over the decades, I accumulated hundreds of stories of funding. One that beautifully illustrates how we survived and thrived came on a day in the early 1980s when our plantation work had stalled for lack of funds. We were short of just ten rupees to buy a bundle of jute rope to tie our saplings. The volunteers were waiting, the sun was fierce and the plants had already begun to droop in their gunny sacks. I sat by the roadside, wondering if the day's work would have to be abandoned. Just then, a rickshaw puller stopped nearby to drink water. He looked at our saplings, smiled and asked what we were doing. When I told him, he nodded, pulled out a folded ten-rupee note from his pocket and pressed it into my hand. '*Yeh ped lagao, Baba*,' he said softly. '*Mujhe toh rickshaw khinchna hai, aap thoda saaya bana do* (Plant the tree, Baba. I have to pull this rickshaw every day, at least make me a shade).'

That ten-rupee note changed something in me. It wasn't about the money; it was about the invisible economy of faith that fuels all true work. We bought the rope, planted the saplings and I have often gone back to that stretch of road. The trees stand tall now, offering shade to countless others who may never know the story of their origin. That day taught me that fundraising is not about collecting money. It is about collecting trust. The shade of those trees is the interest earned on that one small investment of belief.

In 2017, when HCL Foundation came forward to support our work, I felt both joy and disbelief. For forty years, we had been powered by trust, tiffins and ten-rupee notes. Suddenly, there was a formal grant letter, typed on letterhead, stamped, signed and delivered. It felt almost unreal, like nature herself had written it. I remember sitting quietly with that document in my hand, thinking of all the small donors who had carried us this far—the rickshaw pullers, the shopkeepers, the chai vendors. Their contributions had no receipts, no audit trails, only blessings. This letter was their collective signature, proof that the world had finally noticed their invisible partnership.

When the first instalment arrived, our volunteers laughed. 'Baba, now you're corporate!' one of them joked. I smiled and said, 'No, we are still compost.' The money was not a luxury; it was nourishment, like manure, meant to be spread evenly so that new life could grow. We used it for fencing, irrigation and nursery development, ensuring that not a single rupee lost its purpose. That first institutional funding was not a departure from our journey of faith; it was its recognition. It showed that a movement born in love and sustained by trust can eventually be validated by structure, without losing its soul.

Even today, whenever a cheque or a grant comes in, I think of that rickshaw puller's ten rupees. The currency may have changed, but the energy remains the same—belief in action, trust turning into trees.

The Invisible Forest of Faith

As I think back over my life, I realize that our real forest is not just made of trunks, roots and leaves; it is made of people. It is an invisible forest of faith, where every act of kindness is a seed, every rupee a root, every word of encouragement a ray of light. Over the years, these unseen hands and unheard prayers have grown around us like sheltering trees. Some gave coins, some gave cows, some gave corporate cheques, but in truth, they all gave trust.

That trust became our soil. Money can buy saplings, but it cannot buy faith. Shade cannot be purchased; it must be earned through sincerity and service. Our forests, like our friendships, have grown without contracts, watered by belief, nourished by goodwill and protected by time.

Looking back, I realize that every tree planted, every bucket of compost turned, every litre of water poured was not funded by money alone; it was funded by trust, faith and love. That is the real wealth of fieldwork. And so, when people ask me how we survived, I smile and say: 'We didn't just grow trees. We grew trust. And trust, once rooted, becomes the most fertile forest of all.'

Chapter 31

The Trees That Travel with You

The Indian Railways, with their smoky steam engines, were my first moving classrooms. Each transfer of my father, a young major in the Army Medical Corps, became a new syllabus. Allahabad to Pune, Pune to Mumbai, Mumbai to Ahmedabad, through Ratlam, Agra, Mathura, the Chambal jungles, Satna, Morena—every landscape etched itself into my imagination. I learned the country station by station, tunnel by tunnel and platform by platform. The story of India can be read from a train window. So can the story of my life.

For as long as I can remember, I have been obsessed with the trees that guard our smaller railway stations—those magnificent old beings that have watched generations come and go without ever moving an inch. They are the silent passengers of history. Banyans with arms like elephant trunks, peepals whispering to the wind, gulmohars flaring up like sudden applause in summer, neem trees breathing their medicinal calm into the chaos of platforms. These trees were my first companions in travel, my first teachers of patience. The older and smaller the railway station, the deeper the charm, the more human its heart. Faded blue signboards, hand-painted timetables, tea served in clay cups, the hiss of the pressure cooker at the stall selling hot pooris and aloo—all watched over by a tree older than any passenger alive. Those trees, I always felt, knew the timetable of human nature better than the railways did.

The British, in their rigid sense of order and empire, planted avenues of trees along railway tracks: neem, amaltas, rain tree, tamarind and ficus. What began as a utilitarian act, shade for workers, control of soil erosion and stability for embankments, became, over decades, India's longest living corridor of green. If you have ever leaned out of a window between Bhusawal and Jabalpur, you'll see them, rank upon rank of old trees marching beside the rails. They were there before Independence and remain still, unfazed by diesel fumes, indifferent to electric wires. They are the true permanent residents of the Indian Railways.

Each journey was a lesson in geography. From the Deccan's laterite soils to the Gangetic plains, the language of trees changed. Tamarind gave way to neem, neem to peepal, peepal to mango. And somewhere along the curve of the track, the child in me learnt that trees too have dialects. When I later became a tree planter by profession and calling, I realized that railway tracks had always been another classroom. It was the trees that ran beside the rails that taught me continuity, patience and rooted movement.

The first trains I remember were pulled by steam engines—black, loud and alive. They breathed like animals. They roared, they sweated, they sighed. You could smell their breath from half a kilometre away: a mix of coal, iron, grease and adventure.

My father's postings meant that we were always in transit. Dehradun, Dalhousie, Kolkata, Pune, Jalandhar: the names of our stations became the chapters of my childhood. I learnt to read not from storybooks, but from signboards written in three languages.

We always travelled first-class sleeper, with wooden panels, white sheets and brass fittings that glowed faintly in the lamplight. Granny would settle near the window with her flask of tea, and I would stick my face to the glass until my breath fogged the pane. 'These engines have souls,' she once said, 'and they love attention. Look how proudly they puff!' I never forgot that line.

No modern railway terminal can match the personality of the small-town stations of India. I loved their modesty, the iron benches polished by decades of waiting, the ancient fans turning lazily, the single bookstall with Hindi novels and pulp magazines. The smaller the station, the warmer the air. There was one in particular. Guna Junction in Madhya Pradesh, where a grand banyan stood at the edge of the platform. Its roots had grown so thick that they formed their own benches. I often waited there, sometimes for hours, scribbling diary notes. Beneath that tree, I felt a sense of belonging I have never quite found in cities. Under its shade sat the whole miniature India: the tea seller with his rhythm, the cobbler with his patience, the woman selling roasted gram in paper cones. A small bookstall sold the same six titles for twenty years—Premchand, Sarat Chandra and sometimes a Hindi translation of *Jane Eyre*.

The trees watched over all of us. They were the station masters of another order, keepers of memory, guardians of heat and humanity.

Passengers and Pilgrims

In those days, people on trains were not commuters; they were pilgrims of purpose. Everyone was travelling *for something*. To study, to serve, to return home, to chase hope. Compartments were rolling classrooms of sociology. The second-class coaches were my favourites. You could not remain a stranger for long. The moment the tiffin boxes opened, the walls between languages dissolved. A single family's food became a shared buffet. Someone always offered you tea, someone always guarded your bag when you stepped out, someone always told you the next station was still an hour away, so don't worry. India's unity was not written in the Constitution back then. It was baked into railway food, brewed into railway tea and spoken through railway kindness. I have seen families form and dissolve between one junction and the next. I have seen quarrels that ended in laughter and silence that ended in song. These were lessons in empathy that no classroom could offer.

Granny turned every railway station into a laboratory for human study. She would say, 'Observation is education. Don't waste time being bored; look, and you will learn.' So, I looked at the porters and their red shirts faded to pink; the stray dogs sleeping under benches; young ticket collector pretending to be strict but secretly enjoying the attention; vendors calling out in poetry—*chai garam, chai garam!* She made me write in my diary what I saw, how people moved, how trees stood, how dust swirled. She said that the difference between a tourist and a traveller was observation. That lesson would later become the foundation of my ecological work, because a tree planter too must first be an observer of soil, of light, of mood.

When the train moved again, she would whisper, 'Now look at the trees, they'll tell you where you are.' And indeed, they did. Banyans meant the plains. Sal trees meant we were nearing forests. Palms meant the coast. My geography was botanical.

Years later, when I began planting trees across India, I often thought back to those railway journeys. The Indian Railways, without intending to, had created one of the world's longest ecological corridors. From Kanyakumari to Katra, trees line the tracks like green stitches binding the subcontinent together. These railway plantations prevent soil erosion, control dust, absorb carbon and host millions of birds and insects. Every 10 kilometres, there's a small biodiversity miracle unfolding quietly behind the roar of wheels.

The railway lines are heat regulators as well. Temperatures near shaded stations are a degree or two lower than nearby open fields. I have measured it myself. I once joked with a forest officer that the Railways may have done more for India's environment than all our departments combined, simply by accident. When I began leading plantation drives, we worked closely with rail authorities in certain states to regreen abandoned sidings and unused land. The results were astonishing. Within two years, birds returned, shade appeared and the vendors thrived again. A single line of peepals could transform a platform into a community.

Every frequent traveller develops private friendships with certain trees. I have mine. Between Jhansi and Bina, there stands a row of tamarind trees that seem to wave each time the train passes. Once, during a signal halt, I stepped down and touched one. It was warm, as though it had been breathing with the train itself. At Itarsi Junction, there was a mighty peepal beside a bookstall. I bought a novel there in the early 1980s, never finished it, but the memory of reading under that tree's dappled light has outlasted the plot. Decades later, I visited again. The bookstall had new owners, but the tree was the same: thicker, calmer, breaking the platform tiles like an old saint unconcerned with municipal discipline. Such trees remind me that permanence is possible even in a country as restless as ours.

The Waiting Game

Trains are excellent teachers of patience. I have waited six, sometimes eight hours, for trains that never arrived when promised. Yet I do not remember being angry even once. I learnt early that time slows down for those who can watch. Children flying kites on the tracks, vendors sharing jokes, monkeys raiding dustbins, each delay was a story unfolding. The loudspeakers would keep announcing, 'Train delayed by thirty minutes,' until the words lost meaning. By the third or fourth announcement, you began to hear the sound of your own breathing more clearly. That was my first meditation practice. Sitting under a neem tree, chai in hand, no hurry, no phone, no anxiety; just a man waiting with acceptance. That kind of waiting builds patience deeper than yoga.

Trains are moving laboratories of human behaviour. I learnt to read people there the way a forester reads the weather. Crowds have moods. At a missed connection, they grow restless; at the first sight of an approaching headlight, they become hopeful.

Once, on a crowded train in Bihar, a man fainted from the heat. Instantly, ten passengers worked like a single organism. One opened his water bottle, another fanned him with a newspaper, a third offered sugar and a fourth held his head. There was no coordination, no order, just compassion. That is the invisible syllabus of the Indian Railways. Such scenes built my faith in the community, the same faith that later powered the *Give Me Trees* movement. Human goodness, like roots, spreads unseen.

In later years, when I began organizing volunteer campaigns, I dreamt of reviving the old friendship between trains and trees. Railways have miles of unused land, perfect for plantations. Linear stretches where saplings can grow undisturbed, forming green tunnels of continuity.

In Uttar Pradesh and Haryana, our volunteers began small experiments along rural halts. We planted neem, jamun and amaltas. Within a few seasons, the chai stalls beneath them were doing better business. Shade, I realized, was the invisible currency of survival. One station master told me, 'Earlier, no one waited here in summer. Now, people come early just to sit under your trees.' That, to me, was progress, more meaningful than statistics.

If you travel today through certain sections near Yamunanagar or Aligarh, you might still see some of those peepals and neems standing proudly. They are part of the same living corridor that began long before I was born.

Travel has changed. Air-conditioned silence has replaced conversation. Earphones have replaced strangers. But I still prefer the slow trains, the ones that stop at forgotten stations, where trees still shade the water taps and the chai tastes faintly of earth.

Modernity may move faster, but it rarely observes. I miss the smoky engines, the poetry of motion, the sense of shared destiny. I miss the way people looked out of windows, and not into screens.

Trains taught me that life is not about reaching early; it is about travelling fully. They also taught me that roots and journeys are not opposites. You can be rooted like a tree and still travel endlessly, through ideas, memories and connections.

Today, when I plant a tree beside a road, a river or a railway track, I often think of those early journeys. I imagine the children who will lean out of future train windows and wave at these saplings when they've grown. Those trees will outlive the trains, the passengers and even the planners. They will continue to stand witness, to the sound of steel and song, to the faces of a million travellers chasing home. The trees will remember us long after we forget the journeys. Because, in truth, the Indian Railways never just carried passengers. It carried memories, morals and shade. And somewhere along those tracks, I became who I am—a man who plants trees for travellers he will never meet.

Chapter 32

My Scooter Diaries

The scooter was never a machine. It was a friend who understood silence. It had no roar, no race, no arrogance of horsepower, only a hum, a steady breath that matched my own. Together, we travelled not for speed, but for stillness. We were two pilgrims disguised as commuters.

My first scooter was a Bajaj Priya 150. It was light blue in colour, just like the first colour in the inaugural model of the Maruti 800. Then followed a series of Kinetic Honda scooters that came as donations from friends and followers. Volunteers collected money to buy scooters for our plantation teams. They were primarily used as maintenance vehicles. Most of the scooters I have used have travelled over one lakh kilometres each. When the workshop mechanics started warning us not to use the particular scooters on open highways, it was only then that we exchanged them for new models. Even today, I get loads of discounts and excellent exchange offers from our vehicle dealers. They all know the purpose of these scooters. They always tell me that the discounts are their contribution to planet Earth.

It began in Pune, where tar still smelled new after the monsoon. I tied saplings in jute, packed a tin of water and set out without maps, trusting that the road, like an old storyteller, would know where to lead. Soon the city dissolved into open country: sugarcane rustling like gossip, women carrying grass bundles on their heads, men sipping tea under banyan shade. The air changed its accent every few kilometres, and so did the landscape.

The scooter was modest but miraculous. It climbed hills like a mountain goat and rolled down valleys like a tired saint. It never complained, even when loaded with peepal and neem, manure tins and tools and my unreasonable dreams. Its seat was cracked, its mirror half blind, but its spirit was whole. Every day began with curiosity, not an itinerary. I rode through Kolhapur's cane fields, Satara's wind and Nashik's vineyards. I stopped at roadside shrines, poured a little water on thirsty saplings and moved on. Sometimes, when the road was empty and the wind warm, I sang to myself. The engine joined softly, a duet between man and machine.

I learned early that travel is best done without ambition. If you plan too much, you miss the poetry. I slept where the night found me—dhabas, temple courtyards, bus stands or under the old guardians themselves: the peepal and the neem. Their roots held the earth; their shadows held me. Lying beneath them, I often thought about how small our worries look from the ground.

At dawn, tea arrived in tin tumblers, steaming like faith. Truck drivers discussed God, diesel and destiny with equal intensity. Dhabas were my classrooms. The cooks were philosophers who stirred dal with one hand and cosmic truth with the other. One owner near Satara had planted a gulmohar tree outside his stall. He would say, '*Pehle paudha, phir paratha* (First the plant, then the bread).' I never forgot that order of priorities.

From Maharashtra, I drifted north to Indore and Ujjain, where mustard fields glowed like lamps. At every stop, someone would ask, 'Where are you going?' I'd shrug and say, 'Where the road ends.' That answer always made them smile. India loves wanderers; we are a nation raised on journeys, from yatras to migrations, from pilgrimages to work shifts.

Rajasthan tested my endurance. The air there was like sandpaper, and the sun an interrogator. But the desert welcomed me with silence. Villagers offered water as though it were gold. I planted hardy Khejri saplings where even rain hesitated to arrive. 'You're planting in stone,' they laughed. 'Yes,' I said, 'but clouds remember kindness.'

At night, I slept near ponds, lulled by frogs and the faraway moan of trains. Once, a camel stood watching me wake, its expression perfectly philosophical. Rajasthan teaches you humour; it is the only way to survive both heat and fate.

The road to Haridwar during the monsoon was baptism. Rain fell in silver threads, drumming on the scooter's metal body like tabla beats. Pilgrims walked barefoot, chanting, the Ganga swollen and fierce beside them. I stopped near the river and removed my helmet. The rain touched my face with the tenderness of a blessing. For a moment, I felt forgiven for every hurry I'd ever lived in.

Further north, the mountains rose, not politely, but commandingly. The scooter wheezed and panted, but kept climbing. Above Rishikesh, the road wound like prayer beads between cliffs and clouds. At one bend, a gust nearly toppled me. I laughed and shouted into the mist, 'Steady, old friend, even Gods test balance!' The scooter hummed back, loyal as ever.

Through the heart of India, I passed forests where silence had weight, Madhya Pradesh, Odisha, Chhattisgarh—lands where trees still spoke the language of rain. Women carried firewood on their heads and smiled as I rode by, as if recognizing another worker of the Earth. Children waved from the fields. I waved back, sometimes stopping to plant a small tree near their school.

Odisha's coast was a long sigh of salt. Near Chilika Lake, flamingos moved like brushstrokes on water. A fisherman stopped me. 'You catch rain?' he asked, pointing to my saplings. 'Yes,' I said. 'Then we are brothers,' he smiled. 'I catch fish, you catch clouds.' We shared a meal of rice and salt fish, and he sent me off with a blessing as old as the tides.

In Tamil Nadu, temples rose from the red earth like poems carved in stone. Priests tied threads on my wrist and offered me bananas and benedictions. I told one that my yatra was called 'Planting as Prayer'. He nodded and said, 'Then every pit you dig is a shrine.'

The Konkan coast was my favourite rhythm, the road curling between sea and hill, painted in salt, rain and sunlight. Coconut palms leaned as if bowing to greet. In Ratnagiri, fishermen mending nets waved.

In Kerala, it rained the way poetry falls on tired hearts, endlessly, tenderly. The scooter wore a second skin of moss and mud. Rivers crossed the roads as casually as conversations. I stopped often, not for rest, but for wonder. The world looked newly washed.

Heading west again through Karnataka's forests, I felt the road itself breathe. Coffee blossoms scented the air. Monkeys watched from trees, as if guarding ancient secrets. In Sakleshpur, a tea seller told me, 'You are not travelling; you are meditating with motion.' I smiled. He was right.

After a while, travel ceases to be about movement; it becomes about presence.

And so I returned, full circle, to Maharashtra: older, dustier, lighter. My scooter bore the marks of devotion: scratches like scripture, tyres worn to faith. It had tasted salt from three seas and soil from a dozen states. Its engine had heard more languages than any diplomat. Its shadow had sheltered a thousand saplings.

Each journey had been a prayer disguised as travel. I learned to trust the road, the kindness of strangers and the guidance of trees. I learned that a dhaba bed made of jute can give sleep deeper than a hotel mattress, and that hunger dissolves when curiosity is fed. I learned that the true luxury of movement is freedom from fear. If you ask me today what I saw on those rides, I'd say, I saw India breathing—in her forests and deserts, her rivers and mountains, her laughter and labour. I saw the quiet dignity of people who live close to soil and sky. And I saw that the closer you travel to the ground, the clearer life becomes.

Sometimes, when I stand beside that same old scooter, its paint faded, mirrors cracked, keys hanging like prayer beads, I feel it still hum with memory. Each dent has a date. Each scratch has a story. It has carried not just saplings but the scent of monsoon, the dust of deserts and the unbroken rhythm of a heartbeat called hope.

If ever a machine had a soul, mine does. It would whisper, 'We were never just moving. We were growing.' And I would nod. Because somewhere between one dhaba and another, one sapling and another, the road stopped being a route and became a revelation.

Now the scooter waits patiently, like an old monk under the peepal tree, ready for the next sunrise. The key hangs near the door, the helmet gathers dust, waiting for the wind to call me again. When it does, I'll start it with the same faith as before. Because some friendships never retire, and some journeys never end.

This book is a small space to describe three decades of highway adventures. Flat tyres, mechanical failures and total breakdowns in the middle of nowhere will need another book, which I am sure I'll write in the near future. It would be just on my highway adventures across the twenty-two states of India.

According to my diaries, I have covered over 5,40,000 kilometres on roads and highways. This is across 226 districts of India, which I visited to meet volunteers and propagate peepal, neem, jamun and sheesham. Scooters have become like my body extension. I am often asked the question: How did you do it? I just tell them about my scooter journeys. The rest is up to their imagination.

Chapter 33

Bread, Butter and Jam Days

'Between hunger and hope lies a slice of bread.'

—Field saying, 1980s

There are luxury meals and designer diets, and then there is the immortal combination of bread, butter and jam—the fuel that has powered half my life's adventures. It was our breakfast, lunch, dinner, midnight snack and emotional rescue kit, all rolled into one. The menu never changed, but the landscapes did: deserts, valleys, jungles, highways and hilltops. Everywhere, a slice of bread became our passport to survival.

We lived on it for weeks at a stretch. No fire, no ghee, no smoke, no utensils. Just a knife, a jar and a mission. The ground force knew that if the ration was low, at least the bread brigade would march on. It was easy to carry, easy to share and impossible to dislike. You could feed a crowd of twenty in minutes, and nobody would complain, except perhaps a diabetic uncle who'd look at the jam jar with tragic romance.

Sometimes, during planting drives, the team would break for tea under a peepal or neem tree. Out came the bread packets and butter slabs like soldiers reporting for duty. Knives became spatulas, diaries became plates and laughter became the fourth ingredient. The air would fill with the smell of tea, soil and jam, the perfume of a working day well lived.

In every state, from Jammu to Jharkhand, Bengal to Gujarat, this trinity of sustenance followed us like a loyal pet. We have eaten it beside rivers, in sand dunes, at bus stands and once even on a moving tractor. I have seen volunteers wipe jam on their trousers when napkins ran out, and once a cow joined our picnic without invitation. Bread-butter-jam has seen everything—hunger, hope, friendship and foolishness.

Foreign volunteers were often astonished. They came expecting yoga and enlightenment, but what they really got was breakfast enlightenment.

'So this is what keeps India green?' one of them said, holding a sticky slice like a holy relic. I told him, 'Yes, this is our renewable energy source.'

Sometimes, when our numbers swelled into hundreds, we organized what we fondly called *langars of bread-butter-jam*. Supporters brought loaves by the dozens, butter in mountains, jam in rivers. No cooking, no waiting, just the sweet rhythm of spreading and sharing. It was the simplest meal in the world, but it made us feel rich.

Even now, with more resources and better logistics, I keep a loaf and a jar in my bag. The habit never left. Bread-butter-jam is more than food; it is memory, friendship, travel and survival. It has tasted of early mornings in Punjab, late nights in Delhi, cold winds in Himachal and dust in Rajasthan. We owe many forests and many friendships to those slices. They kept us moving when there was nothing else to fall back on. And so, whenever someone asks what kept us alive through forty years of fieldwork, I simply say, 'A good knife, a jar of jam and the courage to spread it wide.'

Those were and still are the bread-butter-jam days—my sweetest adventure partner and my most loyal comrade, my edible poem of survival.

Chapter 34

Room on the Roof

For the past fifteen years, my home has been a *barsati*—a modest rooftop room—in Mayur Vihar, Delhi. That little space, simple and unadorned, became one of the most fulfilling chapters of my life. It was more than a residence; it was a sanctuary where friendships blossomed, conversations flowed and solitude ripened into wisdom. The barsati carried its own freedom. People felt at ease visiting me there. They would call ahead, and I would welcome them as if that rooftop room were a temple of shared thoughts.

My father's life as an army officer had prepared me for this freedom. From him I inherited resilience and adaptability, the quiet art of sleeping anywhere, eating whatever came my way and crossing geographies without complaint. That mobility became my power. The barsati was never confinement; it was a launchpad, reminding me that strength lies in carrying one's world lightly.

Over the years, I discovered that true power comes not from possession but from movement, from belonging in many places, being at ease under different skies. The barsati gave me that gift daily: a view of the city, a small terrace for plants and the sky itself as my constant companion.

From that rooftop, I watched the city change, as it became intoxicated with consumption. New amusements swept across India—DJs, dance parties, carnivals—guilt-free indulgences that became the new normal. Billboards rose like Gods over traffic dictating not only what people should buy but how they should dream. Festivals once lit with the soft glow of clay lamps blazed with electric strings gulping megawatts. Holi, once buckets of water and friendly smears of colour, turned into packaged events with DJs and imported powders. Weddings became competitions of spectacle, fireworks lighting the night while debts silently lit up family ledgers. I sat on my terrace and wondered whether this was progress or only a distraction. Progress should lighten burdens; instead, I saw people staggering under new ones. Consumerism glittered in neon but dimmed the stars.

Even religion, once the quiet binding of faith, grew into the most profitable industry of all, outpacing technology, real estate and even finance. Its product, belief, was invisible, its customers endless, its waste immense and its marketing cleverer than any perfume or car. The irony struck me often: temples echoed with loudspeakers that silenced the birds, the rituals meant to connect us with the divine were producing mountains of plastic waste that would long outlast the ceremonies they adorned and incense thickened lungs with unmeasured carbon footprints. If we can calculate the carbon cost of a turbine, why not a temple ritual?

I laughed bitterly at the business model: no factory, no inventory, only faith for sale. Yet paradox lurked everywhere. The same superstition that choked rivers with plaster idols also protected peepal trees from the axe. Shrines and fears gave them shelter where science failed. Religion polluted and preserved at once, a contradiction that mirrored life itself.

From my barsati, I also watched how politics entered the domain of rivers. Ganga and Yamuna, once sacred presences, were pulled onto political stages as tools of rhetoric. From my rooftop, I could not see the Yamuna, but I could smell her sorrow on certain winds, a dampness that clung like memory. It had become more chemical than water, more froth than fish. While politicians thundered about cleaning, the rivers themselves remained polluted, suffocated and dying. Committees and commissions sprouted like weeds, funds in the billions were announced with fanfare and spent, yet the result remained unchanged: falling water levels, idols resurfacing broken, vanishing vegetation, plastic bags floating among marigold garlands, a toxic foam mocking our prayers and floods where greed had blocked natural flow. Soil, once fertile, had been hardened by decades of chemicals until it resembled plastic. The sponge of the earth was gone. When the rains came, water could no longer sink to replenish the aquifers; it rushed across hardened fields, flooding villages, carrying away both soil and seed. These patterns, visible to every farmer, remained invisible in the corridors of power.

What I learned from that barsati, and from all my journeys, is that change cannot come from policy papers alone. The greatest revolutions, whether industrial, social or cultural, were born from the ground up. Seeds crave soil, not speeches. Roots seek space, not resolutions. Grassroots workers, not think tanks, carry revolutions on their backs. The rooftop taught me this as clearly as the fields did: perspective comes from being both above and within the crowd.

Each season on the roof became a teacher. Those rooftop hours became my sanctuary, a place where I was both a participant in the city's frenzy and a witness standing apart. My barsati was a sanctuary for writing. Surrounded by plants and visited by friends, it was a place where thoughts clarified. Lying on the cracked terrace, I would ask questions not to receive answers but to be emptied of noise. What is freedom? What is love? Why this endless striving? The answers came not in sentences but in clarity, in whispers that only silence could hear. Many of my best insights came while sitting at a small desk beneath a leaking roof, listening to the city hum below and pigeons cooing above. I discovered that writing was not just a record but a rhythm. One must show up, again and again, to meet the words halfway.

Fifteen years in a room on the roof—years that have been adventurous, grounding and full of companionship. The barsati was not just my home; it was my teacher. It taught resilience, perspective and gratitude for the simple joy of having the sky for a ceiling illuminated by stars burning with indifference, while Delhi roared with markets and machines. It provided the balance I needed to remember that beneath all our frenzy, the Earth still turned, silence still endured, the sky still held its vast calm.

Summer

Delhi summers on the rooftop were like living inside a furnace, yet they became chapters of growth. By midday, the walls radiated heat so fiercely that touching them felt like pressing a hand against a stove. The ceiling fan, loyal but lazy, merely sighed instead of spinning, like an old soldier tired of battle. Those were the months I learned to surrender rather than resist. I moved my cot onto the terrace with a pillow, a thin sheet and a bottle of water. Lying under the naked sky, I stared at constellations flickering faintly through layers of dust and smog. Those nights were uncomfortable, yes, but magical too. The air was heavy, but my mind grew lighter with every hour spent gazing upward. Sometimes I traced the path of Orion; sometimes I followed the slow crawl of a blinking satellite. At other times, I simply listened to the city's nocturnal orchestra, barking dogs, the night guard's whistle and a distant train. Every sound seemed magnified on the roof, yet softened by distance.

Around ten, families on neighbouring terraces spread out their beds, children giggling as they settled down. By midnight, silence descended, broken only by the clatter of utensils or a sleepless cough. The air cooled slightly before dawn, and in that fragile hour, I wrote in my diary by the light of a kerosene lamp. Thoughts that felt heavy in the day flowed more easily under the stars. Summer, harsh as it was, gave me patience and taught that discomfort, too, can be a kind of companionship.

Monsoon

If summer was fire, the monsoon was theatre—grand, unpredictable and drenched in laughter. The first clouds of the season arrived with a whisper of excitement, gathering from the west until the sky turned from blue to an ocean of grey. Thunder rolled across the Yamuna plains like a drum announcing the arrival of Gods. And then the rain came—fierce, unapologetic, hammering against the tin shade above, pouring down the staircase, flooding the terrace in minutes.

Inside, I scrambled with buckets and pots, catching leaks from the cracked ceiling. Each vessel filled quickly, water sloshing as I replaced it with another. My barsati transformed into an orchestra of droplets—*plink, plonk, splash*—each note echoing off the walls. I often laughed at the absurdity of it, sitting cross-legged on the damp floor, a candle flickering beside me when the power inevitably went out. 'An amphibious dwelling,' I joked to myself, 'perfect for a tree planter.'

The smell of wet earth rose in waves, mixing with jasmine and neem from the neighbourhood. Books curled at the edges, and more than one diary succumbed to fungus, ink bleeding like memories refusing to stay contained. Yet the monsoon never felt cruel. It was mischievous, demanding, insistent, but also generous. When the rains stopped, the terrace gleamed under a rinsed sky. Frogs croaked from hidden corners, and children danced barefoot in the lanes below. The monsoon taught me to surrender, that water always finds its way, and sometimes wisdom lies only in letting it.

Winter

Winters on the rooftop were the quietest of all. Fog rolled in like a soft curtain, muffling the city's usual clamour. The air bit at my fingers, turning my breath into little clouds. I wrapped myself in layer after layer—sweaters, mufflers, socks—and still felt the chill seeping through. Yet in that stillness, reflection arrived.

The terrace at night was a place of lantern light and silence. I often sat wrapped in a blanket, notebook on my lap, scribbling until my hands grew stiff. The fog made the city mysterious: buildings vanished into white, lights became halos, sounds softened into whispers. From the rooftop, Delhi seemed a dream half-remembered.

Winters were also for companionship. Friends came wrapped in shawls, carrying roasted peanuts, and we sat huddled around steaming cups of tea. Stories flowed more easily in the cold, perhaps because warmth is shared not just through fire but through words. Sometimes silence was enough. We would sit, watching fog drift past, knowing the moment itself was sufficient. Winter taught introspection. If summer was endurance and monsoon surrender, winter was reflection—the time to gather lessons and quietly prepare for renewal.

Spring

Spring in Delhi was brief, but the barsati rejoiced in it. Flowers bloomed in the pots—marigolds bursting with gold, hibiscus flaming red, bougainvillaea cascading pink over the parapet. Butterflies fluttered in morning light; bees hummed lazily; the terrace turned into a hanging garden above the city.

The air softened, carrying the scent of mango blossoms and mustard fields from across the Yamuna. Children flew kites in the afternoons, their laughter rising as colours danced against the sky. Friends came to bask in the sunlight, shawls draped loosely, cups of chai warming their hands. We spoke of dreams, of small joys, of plans that may or may not come true. Spring invited hope, and hope came easily. Branches that seemed lifeless in winter suddenly sprouted green. The city, too, seemed to breathe again. Spring whispered that renewal is inevitable: after fire, after flood, after fog comes blossom.

The barsati, though simple and bare, became a gathering place for countless souls. The staircase that led up to it carried more stories than its cracked steps could hold. Over the years, I welcomed friends, colleagues, volunteers, seekers and strangers who had heard of the quiet rooftop where tea was always brewing. Each visit added a new thread to the fabric of my life.

Some afternoons, volunteers from plantation drives climbed up, dusty and smiling, their hands still smelling of soil. We sat cross-legged on the floor, gulping water before shifting to steaming cups of chai. Conversations began with logistics, how many saplings survived and where water was scarce, but soon wandered into philosophy.

One young volunteer once sighed, 'I thought we were saving trees, but I realize now that the trees are saving us.' His words echoed long after the evening ended.

Many treated the barsati as a confessional. The absence of luxury made it easier to unburden the heart. Visitors came with stories of heartbreak, family feuds or crises of faith. I listened, not as a counsellor but as a fellow traveller.

Sometimes all a person needs is silence in which their words can land softly.

Once, a colleague arrived with eyes heavy from unshed tears. She spoke of feeling lost in the city, unseen despite the crowds. We sat on the parapet, watching the sky turn from gold to indigo. When words ran out, silence took over, and she later told me that silence had healed more than any advice could. The rooftop became that kind of space, where silence was sacred, not awkward.

Chai was the heart of it all. I brewed it in a dented aluminium pot, strong with cardamom and, in winter, a touch of ginger. Clay cups carried their fragrance, and the first sip always softened the hardest expressions. I often joked that the tea leaves were wiser than I was, for they seemed to coax truths out of people. Over chai, friends shared dreams, business ideas, journeys and hopes of love. Others confessed fears, loneliness, failure and uncertainty. The rooftop held it all, and chai became the thread binding our conversations.

Once, a group of students filled the terrace with chatter and debate on politics, professors and purpose. By the time the last cup was empty, they agreed on only one thing: that the rooftop felt freer than their classrooms, freer even than the city streets.

Not every gathering was solemn. There was laughter too, the kind that shook the parapet walls. Arguments flared about whether progress was real, whether faith was illusion, whether planting saplings mattered when forests were burning elsewhere. But arguments under the open sky are humble things; they dissolve before ego can harden.

We disagreed, laughed at our disagreements and poured another cup.

Sometimes silence replaced talk altogether. We sat side by side, watching city lights blink, listening to the koel call. On that roof, silence was dialogue, one that spoke directly to the soul.

Not all visitors were expected or called ahead. Neighbours climbed up with bowls of halwa or plates of pakoras, insisting I eat. Children from the lanes below would sneak up to look at the books and plants, their eyes wide with wonder. Once, a street dog found its way up and refused to leave, curling on a mat as though it had always belonged. The rooftop welcomed everyone, human or otherwise.

There was even a wandering sadhu who appeared one evening. He said he had heard of a place of peace on this terrace. We sat together in silence, sipping tea. Before leaving, he smiled and said, 'Peace walks quietly wherever simplicity lives.' His words became etched into the damp walls, joining countless echoes already stored there.

Looking back, I realize that the rooftop was less a room and more a mirror. It reflected to each visitor what they most needed to see: hope, relief, clarity or companionship. For me, it was proof that wisdom does not always descend from mountains or flow from ashrams. Sometimes it brews in a dented pot, served steaming in clay cups, under an honest sky.

The rooftop taught me that every conversation is sacred if held with sincerity, and every silence is sacred too, if shared with presence. Those evenings of chai and laughter and confessions and arguments remain some of my richest memories. They stitched people into my life one story at a time and turned that barsati into a sanctuary, not just for me, but for all who came.

A Balcony to History

The rooftop was my balcony to history in motion. From its parapet, I watched Delhi transform, season by season, festival by festival. There were mornings when the horizon was clear, and I could see cranes stretching skyward, planting their own metallic forest. By evening, new advertisements unfurled across billboards, neon signs blinked into life, and the skyline grew more crowded.

Where open fields once lay, malls rose like temples of glass and steel. The humble corner shop gave way to supermarkets. Where children had played cricket on dusty lots, parking garages swallowed the ground. It was dizzying to watch. The city's rhythm quickened, as if everyone were running toward something they could not name.

From the barsati, I saw both the charm and the cost. Families in their best clothes visited malls; children marvelled at escalators; teenagers basked in air-conditioned theatres. Progress was undeniable. But so was the silence of sparrows that no longer nested, the smog that thickened each winter, the noise that drowned the koel's song.

Delhi became more than an address. It was a classroom, a testing ground for dreams slowly turning real. That one-room barsati was a laboratory of change. From there, countless plantation plans were drafted on kitchen tables. Maps of roads, canals and city parks covered the walls. My scooter waited outside, ready for its daily duty, carrying saplings, manure and ideas through Delhi's choking traffic.

Each evening, as the sun sank behind the smog-wrapped skyline, I watered the pots on my balcony—tulsi, neem and peepal—and thought of the trees waiting beyond the city's edges. Delhi was bursting at the seams, yet in its cracks I saw possibilities. The city was not a burden; it was a challenge. If green could survive here, it could survive anywhere.

In the years that followed, our small, ragtag team spread like roots underground. Highways, canals, campuses, colonies, every patch of land became a nursery. Citizens joined in, schools opened their gates and corporations began calling for advice. *Give Me Trees* had become a collective heartbeat for Delhi, a living forest in motion.

Sometimes, when I return to that old lane in Mayur Vihar, I slow down by the gate and smile. I can almost see my younger self, cross-legged on the floor, surrounded by notebooks, scribbling late into the night, unaware that those notes would one day grow into something vast and beautiful. That small room taught me the greatest lessons: how to live with less, listen more and turn a seed of thought into a field of action. The journey from that single room to hundreds of plantation sites across the country has been nothing short of miraculous. Every sapling planted since carries a memory of that room, its silence, its struggle and its steady faith. And perhaps that is what my barsati, like every rooftop, truly is—a halfway place between earth and sky, between what roots you and what sets you free. In the end, every great forest begins the same way: with one seed, one hand and one quiet room on the roof.

At night, when the city folds itself into silence, the rooftop still remembers. It remembers the laughter that rose like smoke, the clink of cups, the murmur of rain, the slow writing of dreams. The wind turns the pages of old diaries, and sometimes, I imagine the plants whispering to one another, stories of summers survived and monsoons forgiven. Up there, beneath the slow pulse of distant stars, I learned that peace is not found in grand escapes, but in small, honest spaces. Perhaps heaven was never far away; it was always waiting, quietly, in a room on the roof.

I am writing this book in this barsati, a humble rooftop room that has been my home, office and ashram for years. From here, among tin roofs, pigeons and potted plants, we have built and nurtured an internationally acclaimed ecological restoration movement. The world may imagine grand offices with glass walls and conference tables, but our headquarters has always been this little terrace room in Mayur Vihar. It leaks in the monsoon, bakes in summer and freezes in winter, yet it has witnessed some of the most meaningful conversations, friendships and revolutions of my life. It is amazing how a small barsati on a rooftop in East Delhi became the nerve centre of a movement that now breathes through thousands of volunteers across India. This is not a story of comfort; it is a story of conviction. A story of how love and faith, when rooted deep enough, can grow forests even from the cracks of a concrete city.

Chapter 35

Spirituality

Spirituality is not something I found in a temple, nor did I inherit it from any scripture. It came to me slowly, like dew forming on leaves—quiet, invisible, inevitable. It came through pain, and through the long, patient process of learning how to come out of it.

I have seen suffering in many forms—the silent kind that hides behind a smile, and the loud kind that shatters through the heart like lightning. In the early years, I believed suffering was to be fought, resisted and conquered. Later, I learnt that suffering is to be understood, not defeated. It is a teacher, and it stays until the lesson is complete.

Spirituality, I have realized, is a skill, not a sermon. It is the art of stepping out of your own darkness without blaming the night. It is knowing where the door is, and having the strength to walk through it again and again. Most people learn how to enter pain; very few learn how to exit it gracefully. The trees taught me this. A peepal tree loses all its leaves in summer, stands stripped and bare, and yet never complains. It knows the shade will return. It has faith in its own roots. That faith, that quiet, trembling, unseen strength, is spirituality.

To stay steady in one's truth, to remain anchored while the world keeps changing its definitions of success and failure—that is the mark of a realized soul. I have met people who meditate for years but lose their peace at the sound of a honking car. And I have met old gardeners who never spoke of meditation but lived in a state of constant prayer, their fingers in the soil, their eyes in the sky. Realization is not about renouncing the world; it is about being so completely part of it that you no longer see a boundary between you and the rest. You breathe for the trees, and they breathe for you. You water the soil, and the soil remembers.

When I realized this, I stopped seeking God in shrines and found him in saplings. Each seed became a silent scripture, each drop of water a mantra. I understood that enlightenment was not about light descending from above; it was about realizing that the light was always inside, waiting to be uncovered by compassion, patience and clarity.

There comes a time when words become unnecessary. When your very being starts to teach, not by effort, but by presence. A lamp does not announce that it gives light; it just does. That is enlightenment.

When people tell me that they began planting after meeting me, I do not think it is because of anything I said. Words fade. But the vibration of a life lived in surrender—that travels. It reaches others silently, like the fragrance of a flower carried by the wind.

Enlightenment, I believe, is when you stop needing recognition for the good you do, and when you stop needing company in your solitude. You become one with the rhythm of existence, the rain, the decay, the bloom, the silence. You understand that every leaf is born only to fall, and yet, in its brief existence, it performs a sacred duty to feed life.

In the end, I have realized that spirituality is not an achievement. It is an understanding, a soft, fragrant awareness that passes through your being and makes everything sacred. To be spiritual is to learn how to walk out of your own suffering. To be realized is to stay centred in your peace. To be enlightened is to help others, wordlessly, effortlessly, by simply being who you are. And if, in the quiet of your heart, a few trees grow because of your presence, then perhaps that is all the enlightenment the Earth ever asked of you.

Chapter 36

Learnings from the Desert Continent

In a land where rain is shy and soil is stubborn, I found lessons in patience, precision and quiet care. Australia, the desert continent, taught me that greenery is not a gift from the skies but an act of will. If they can make deserts bloom, we can surely make our rivers remember how to sing. Australia was never meant to be this green. It is a continent of dust, drought and discipline. The soil cracks, the rivers falter, the winds arrive uninvited and yet, the land endures. If they can coax life from rock and salt, why can't we—blessed with monsoon, mountain and meadow—treat our abundance with the same devotion?

To understand Australia's greenery, one must begin with its geology. It is one of the oldest landmasses on Earth and its rocks whisper stories more than three billion years old. The Australian Shield, a geological feature made of molten rock that has cooled and solidified over half the Australian continent, has not been folded or uplifted for ages; it lies bare, ancient, eroded and silent. The Great Dividing Range is not a mountain range in the dramatic sense of the Himalayas; it is an old, weary rise of stone that has seen oceans form and vanish.

Its soils are thin, tired and nutrient-poor, among the oldest and least fertile on the planet. Ironstone, sandstone and laterite dominate. Rivers wander slowly, often without finding the sea. Water sits in subterranean basins, especially the Great Artesian Basin, the largest underground reservoir on Earth. In many parts of the outback, it rains once a year, or not at all. And yet, this ancient land supports a symphony of life so unique that 80 per cent of its species are found nowhere else.

Australia is a living geology lesson, a museum without walls. It shows what patience can do for survival. Evolution has learned to be frugal. Plants grow not fast, but deep; not broad, but wise. The eucalyptus turns its leaves vertically to avoid the sun. The acacia folds its foliage to save water. Even the soil crust, a thin skin of cyanobacteria, guards every drop of moisture like prayer beads. Standing there, I felt like a student again. Every rock was a teacher. Every leaf was a research paper.

Australia treats nature not as a charity, but as a discipline. From the Great Barrier Reef to the red heart of Uluru, every ecosystem is mapped, monitored and managed. Conservation there involves both science and the social contract. Citizens know the value of compliance, not out of fear, but from faith in systems. Their national parks are not ornamental reserves; they are living laboratories of coexistence. Fire management, invasive species control, seed banks and wildlife corridors are treated with the seriousness we often reserve only for economics.

Australia taught me that ecology thrives on accountability. Rangers there carry tablets instead of rifles. Every tree, every koala, every patch of regenerating scrub is recorded, tracked and revisited. They audit ecosystems the way we audit accounts. Their faith lies not in slogans but in spreadsheets, yet behind the data is devotion.

I walked through dry riverbeds that came alive in the faintest rain, and through forest edges where signage read: *'This area is under regeneration, please let it heal.'* Imagine a country where even a weed receives due process before removal. I realized then: Conservation is not nostalgia; it is governance.

Adelaide, the Quiet Planner

Adelaide is a whisper, not a shout. It does not rush to impress; it unfolds. Clean avenues, generous parks and a rhythm of living that feels contemplative. There, I learned that creativity, whether of pen or plant, needs solitude. A book or a garden cannot grow in noise. City councils are like gardeners in uniform. They layer the soil with compost and mulch before planting. They do not fight the earth; they prepare it. They mix, turn and rest the ground until it welcomes roots. Even their median strips bloom. Adelaide taught me that beauty is built, not bestowed. That silence and structure are two forms of respect.

Perth, the Desert Engineer

Perth sits at the edge of the Indian Ocean, where sand and sea share the horizon. The city is proof that aridity need not mean sterility. At Kings Park, the vast Royal Botanic Garden, I saw resilience turned into an art form. The plants are chosen not for show but for survival: hardy shrubs, native grasses, eucalyptus that seem to drink sunlight. Woodchips, little marvels of life created from the tress themselves, provide what these plants need.

Sydney, the Confident Coexistence

Sydney dazzles. The harbour gleams like liquid glass, the skyline arches with pride, yet the green remains. Here, progress and nature are not adversaries but neighbours. The Royal Botanic Garden flows into the Opera House lawns, the mangroves meet the ferries, and cockatoos gossip in fig trees older than the city itself. Sydney's genius lies in its integration. Storm drains become wetlands. Bridges shelter nests. Coastal marshes are preserved even as towers rise. It's as if every architect in Sydney has a secret second degree in ecology. The people walk briskly but notice leaves. Their politicians cut ribbons for bike lanes, not billboards. Their development doesn't erase rivers; it listens to them.

Melbourne, the Philosopher's Garden

Melbourne is introspection in motion. A city that reads poetry, drinks rain and names its trees. The Royal Botanic Gardens there feel like scripture rendered in green. Under the grand Moreton Bay Fig, I stood silent. Its roots twisted like frozen rivers, its trunk thick as a myth. Beneath it, I realized that the difference between their gardens and ours was not one of wealth, but of reverence. Their gardeners study the soil the way monks study scriptures. They debate the angle of irrigation emitters with the seriousness of philosophers. Melbourne thinks in centuries. Every sapling planted there is an act of patience written into policy. It is a city that believes in shade for unborn children.

If there is one metaphor for Australian environmental philosophy, it is the woodchip: small, silent, essential. Each flake of bark tells a story of circularity: yesterday's tree feeding tomorrow's. It is humility in action, the science of simplicity. Where we burn leaves, they mulch them, spreading them in a thick layer that protects every tree base and nourishes every patch. When we sweep away the fallen, they turn it into a foundation. Their gardening is not decorative; it is devotional. Woodchips hold moisture, prevent weeds, feed fungi and return as soil. To watch them unloading truckloads of mulch, spreading it evenly with calm precision, was to watch a quiet ritual of faith. In Perth, gardening is not a hobby; it is urban engineering wrapped in tenderness.

Their cities are fewer, newer and smaller. Ours are ancient, abundant and defiant. Yet somehow, they breathe easier. Why? Because they plan before they plant. Because their policies have patience built in. Their mayors talk about shade, soil and water; ours talk about slogans. Their budgets list mulch, biodiversity and rainwater harvesting; ours list pavements, concrete and contracts. Their leaders cut ribbons for saplings; ours cut trees for roads. But this is not a complaint; it is a contrast. We have the heart; they have the habit. If we could borrow their discipline without losing our devotion, our cities would bloom differently.

When I returned home, I saw our saplings with new eyes. Survival of plants needs intent. Intent is invisible irrigation. You cannot see it, but the plants feel it. We fought for the life of our baby plants like parents of children in an incubator. Volunteers carried buckets from hand pumps, shielded shoots from goats, and loosened the soil after each monsoon. Where Australia had technology, we had tenderness. Our afforestation science was written not in manuals, but in mud.

We learned that water is not a commodity, it is emotion; that soil is character; that species selection is identity. We planted peepal, neem, banyan and jamun—our native saints of survival. We learned to nurse, not just plant. Survival taught us humility. It taught us that nature rewards patience, not passion. The world runs after speed, but a gardener learns to run after patience. Planting is easy. Survival is sacred.

Sometimes I think, even this book will be dust one day. But like the woodchips, perhaps it will feed something unseen. That is what the desert continent taught me: nothing lasts forever, but everything can nourish what comes next. To plant, to write, to share: each act is a leaf in the great compost of time.

And as I left the continent behind, I knew that the learning had only begun.

Chapter 37

The New Flying Rats

Pigeon love!

Every Indian city today is fluttering with pigeons. From Delhi to Pune, from temple courtyards to traffic lights, we have officially fallen in love with the new flying rats. They have replaced the crow as the national background bird, the squirrel as the morning entertainer and the stray dog as the city's most photographed animal. They are everywhere: in selfies, in temples, on balconies, on statues, even on your freshly washed shirt left to dry in the sun. The blue rock pigeon has conquered our concrete cliffs, and we, the so-called superior species, have turned into their full-time caterers.

In Delhi, science has been pushed out of the nest. There are now over fifteen thousand pigeon feeding stations across NCR. Every traffic island and temple compound has become a sacred restaurant for birds. The business is simple: buy a sack of grain, find a pavement, pay off the local constable and voilà—you are now a licensed distributor of blessings. Tax-free, rent-free, guilt-free. It's the only startup where the product is karma and the customers are both human and avian.

Down south, in cultured Pune, the feeding takes a more disciplined form. Near the Alka Talkies Junction, I once saw a gentleman standing serenely at a traffic signal. He had a large bag of grain and the poise of a yoga instructor. Cars honked, scooters swerved and he remained calm, scattering millet with divine concentration.

'Sir, the pigeons are blocking the signal,' I said. He smiled without looking up. 'Let them eat. They have souls too. 'Yes,' I said, 'but the rats will join the feast.' He frowned. 'Don't insult God's creatures.'

'I'm not insulting,' I said, 'I'm warning. The rats will take it as an invitation.'

He didn't reply, and I didn't insist. One learns early in environmental work that logic rarely wins against sentiment. Especially when the sentiment wears sandalwood paste and carries a bucket.

Meet the Bird Behind the Blessing

Let's talk about science for a minute. The bird everyone calls 'kabootar' is the blue rock pigeon, a descendant of the wild rock dove. Its natural habitat? Sea cliffs and caves. Its natural diet? Seeds, grains and small insects. Its natural predators? Falcons and common sense. Both are missing in the cities.

In nature, pigeons breed a few times a year. In cities, with a constant food supply, they breed all year round. A pair can raise six to eight broods annually, each with two chicks. Multiply that by the thousands of feeding stations and you get a rough estimate of our collective foolishness. Their droppings, rich in uric acid, corrode monuments, paint and marble. Their dried feathers clog air ducts. Their faeces harbour *Cryptococcus neoformans* and *Histoplasma capsulatum*—fungi that can cause serious lung diseases, especially in children. But of course, what are fungi compared to faith?

The Business of Blessings

In Delhi, the pigeon industry runs smoother than the metro. At many feeding points, a local 'manager' sells packets of grain at twenty rupees apiece. Devotees buy, feed, smile and leave, convinced that they've scored cosmic brownie points. The manager collects the day's earnings, shares a portion with the local policeman and sleeps well knowing he's in both business and religion simultaneously.

Astrologers, meanwhile, promote the habit as planetary therapy. 'Feed pigeons for Saturn,' they advise. And people do, generously, religiously, sometimes competitively. You haven't truly achieved urban sainthood until a hundred pigeons land on your head in Connaught Place while someone photographs your halo of feathers.

The irony? Every handful of feed meant for birds feeds corruption as well. Money flows from pocket to pocket, grain from sack to street, and droppings from sky to face. Science stands helpless, sneezing in the corner.

But the saddest part isn't the corruption; it's the ecological eviction. Sparrows, bulbuls, sunbirds and mynas—all have retreated from our balconies. The pigeons are bullies with wings. They claim space, guard it fiercely and drive others away. Their monopoly is absolute. Even crows now look uncertain around them.

And then there are the rats. Oh, the rats! Every feeding station becomes a midnight buffet. Whatever the pigeons drop, the rats mop. Come nightfall, the city's lower floors come alive. It's a perfect partnership: pigeons above, rats below, humans wheezing in the middle. An urban food chain powered by faith and ignorance.

Municipal workers quietly curse this compassion. The droppings clog gutters, corrode railings and turn marble black. Every morning, sweepers begin their day not with a broom but with a sigh. But no politician dares touch the issue. The last time someone tried to regulate stray dogs, the outrage nearly toppled a department. So, pigeons remain protected by superstition—the strongest law in the land.

Doctors, of course, see the fallout. Asthma, bronchitis, hypersensitivity pneumonitis: the hospitals are full of children gasping their way through divine benevolence. But who listens? 'Feeding pigeons brings good luck,' says the pundit. 'Yes,' replies the pulmonologist, 'if you define luck as oxygen cylinders.'

Once, in Pune, I decided to test an experiment of my own. Near a row of pigeon feeding spots, I planted saplings of neem and peepal. Within months, the shade attracted insects. The insects attracted sparrows. The sparrows attracted balance. Slowly, the pigeons moved away, preferring open concrete where the feeders operated. People asked me what I had done. 'Science,' I said. 'And patience.' They nodded respectfully and then, of course, went back to feeding pigeons. Old habits die harder than hungry birds.

Our compassion is admirable; our direction, questionable. We feed because it makes us feel good, not because it does good. We've confused sentiment with service. True kindness, after all, doesn't need an audience. It's quiet, balanced and rooted in understanding. Feeding pigeons on a footpath is like donating sugar at a diabetes camp; the intention is noble, the effect disastrous.

If the government truly cared, it could act smartly. Introduce a Pigeon Feeding Tax. Register every feeding station. Use that money to clean monuments, disinfect ventilation systems and educate the public. But such policies require courage, and courage is not available in government stock these days.

Every city reveals its own truth. Delhi taught me that superstition scales faster than awareness. Pune taught me that politeness can be more persistent than logic. Both taught me that pigeons have done what no politician could—unite the country under one cause: misguided compassion.

Yet, I cannot entirely blame the bird. The blue rock pigeon is simply doing what nature designed it to do: survive. It found in us a species generous enough to feed and foolish enough not to learn. It adapted faster than we evolved. In a way, the pigeon is the perfect urban guru: calm, consistent, shamelessly self-interested and always surrounded by devotees.

Sometimes, while riding my scooter past a row of feeding stations, I see them, hundreds of pigeons pecking greedily, their necks shimmering in the sun like old silverware. People stand around with folded hands, tossing grain and guilt together. A policeman watches, amused. A rat scurries by, collecting leftovers. Somewhere, a sparrow flies past unnoticed, looking for a home.

I sigh, start the engine and ride on, past the fluttering wings, the holy confusion and the haze of good intentions gone wrong. The city hums, the pigeons coo and the lungs wheeze in rhythm.

Until we learn to feed wisely, the sky will remain grey with misplaced mercy. And below it, in every home and hospital, the new pandemic will quietly spread, invisible, polite and shaped like a pigeon feather drifting in the Delhi air.

Chapter 38

The Great Forgetting

There was a time when winters in India had a soul. They arrived gently, like a sigh of relief after months of blazing sun. The mornings were wrapped in mist, and the afternoons were golden with a light so soft it could heal. Smoke from the *chulha*s curled into the air, mingling with the smell of mustard fields. Life slowed down. People paused. Even the earth seemed to breathe differently. Those winters are gone. Today, they come late, leave early and arrive uncertainly, half-hearted, hesitant and exhausted.

The fog that once caressed the fields now suffocates the cities. What we call winter smog is nothing but a graveyard of forgotten promises, of cleaner air, greener roads, wiser choices. Our seasons have blurred into one long, feverish stretch of heat, broken occasionally by violent storms. Summers are unbearable, monsoons unpredictable and winters forgettable. The planet is running a fever, and the thermometer refuses to drop. Every passing year is the hottest on record. We read it in the papers, shake our heads and carry on. We scroll past catastrophe.

The great forgetting has begun. We forget that there were once days when the sunlight was tender, when shadows stretched lazily under trees, when the breeze carried the sound of birds and not engines. We forget the scent of dew on the soil. We forget that the earth was kind once, and that we were, too. We have poisoned everything that once healed us. Our air burns the throat, our rivers carry plastic instead of fish and our soil is thick with chemicals instead of worms. Once we cleansed ourselves in rivers; now the rivers beg to be cleansed of us. We have replaced the fragrance of forests with the stench of exhaust, and we call it progress.

India is burning, not just in heat, but in spirit. The heatwaves come like punishment, turning fields into deserts and dreams into ash. Forest fires rage where cool winds once blew. Crops wither, farmers despair and the cycle of seasons—the sacred rhythm that sustained us—is broken. Our agriculture is crying for rain that does not come, for soil that no longer breathes. The very earth that once gave us abundance now lies cracked, dry and thirsty.

And what about the people? They are burning too. Inside their minds, in their hearts. Mental health is the new drought. Anxiety spreads faster than weeds and depression grows in concrete jungles where no roots can find hold. The body can adapt to heat, but the mind cannot adapt to loss—the loss of coolness, the loss of quiet, the loss of belonging. When you move away from nature, you move away from yourself. When you forget the tree, you forget your own shade.

Our civilization thrives on denial. We built more flyovers while cutting the last standing peepal that shaded the road. We erect glass towers that trap heat and then invent machines to cool them. We pollute the rivers and then bottle water to sell at a price. We cover the soil with concrete and then complain that the ground doesn't absorb rain. The irony is blinding. The tragedy is banal. The madness is systemic. The forgetting is complete.

There is a strange arrogance in believing that we can live without trees, that technology will outwit temperature, that concrete will replace compassion. We forget that the planet doesn't negotiate; it responds. Every chopped tree is an entry in the ledger of heat. Every polluted river is a message written in drought. Every smoggy sunrise is nature's obituary for our wisdom. Yet we continue to act as if we are separate from it all, as if the atmosphere is an inconvenience and the soil a servant.

Our winters are lost, and with them, our sanity. The mornings no longer invite reflection; they demand endurance. The night air no longer soothes; it suffocates. The stars fade behind a haze of human arrogance. We are a species that refuses to learn.

We must at least save what remains. Save the last standing trees, the last clean rivers, the last quiet corners where birds still sing. Save the memory of shade. Save the possibility of coolness. Because once these are gone, no machine, no invention, no prayer will bring them back. We can rebuild cities and industries, but we cannot rebuild seasons. We cannot replant a climate once it collapses. The winters we lost were not just about temperature; they were about temperament. They taught us stillness, patience and introspection. Without them, the year feels incomplete, hurried, breathless.

The human memory is jinxed. It erases pain so we can move forward, but it also erases wisdom so we keep falling back. We forget the flood as soon as the water recedes, the fire as soon as the rain arrives, the pandemic as soon as the traffic resumes. We call this resilience. It is actually amnesia. And it may cost us our planet. Yet I believe it is not too late. The Earth, for all our cruelty, remains forgiving. A single peepal tree can still lower the temperature of its surroundings by several degrees. A grove of neem can clean the air better than any purifier. Shade is the oldest technology we ever had. Each tree spared is a small victory, each seed sown is a small rebellion. The act of planting is the act of remembering. When you plant a tree, you remember that you belong to the soil, not the other way around.

Perhaps one day, the winters will return. Perhaps the fog will descend again over mustard fields, the mornings will hum with life and the evenings will cool our tempers as much as our skins. But until then, the least we can do is not forget. Forgetting is how the loss began; remembering is how healing will start.

The planet does not need saving; we do. And every tree is still willing to help.

Chapter 39

The Forgotten Pandemic

The whole world went silent once. Not symbolically, but literally. Cities fell mute, highways stood empty and railway tracks rusted in the heat of inaction. The hum of the world—that constant background noise of machines, markets and movement—stopped and there was silence across the globe. We were left with the sound of birds, the rustle of trees and our own breathing. For a brief moment in human history, the planet had time to listen to itself and for the first time in decades, we could hear the planet breathe.

And yet, just a few years later, it feels as if it never happened.

I remember those early days of the pandemic not as a pause, but as an awakening. While most people were locked inside, waiting, worrying and watching news tickers count invisible deaths, we were out in the fields. For us, nature did not shut down. Trees do not take lockdowns. Watering cans do not carry curfews. The saplings still needed care, and the young plants still needed water.

We were given special passes by local administrations, thin slips of official paper that allowed us to move while the world stayed still. Those papers became our passports to the living world. Roads were empty, checkpoints quiet. Our teams drove through cities that looked abandoned, but the moment we reached our nurseries, it was as if time still flowed normally. Saplings were growing. Leaves were unfurling. Peepal roots were gripping the earth. The silence of the world made the songs of the forests clearer.

For the first time in decades, there was no interruption, no event, no gathering, no traffic jam, no volunteer confusion. Just pure work. We watered, we mulched, we composted, we spoke to each other less and to the soil more. Every plant became a heartbeat we could still depend upon. The world called it lockdown, but for nature it was liberation. The skies above Delhi turned blue again, a shade we had almost forgotten existed. The Yamuna flowed with a hint of clarity. Birds returned to balconies. Civet cats were spotted on city roads, peacocks danced in colonies and even the shy nilgai appeared near highways as if reclaiming old routes.

Satellite images showed the Himalayas visible from the plains of Punjab for the first time in decades. I received photos from volunteers in Ludhiana, Dehradun and Allahabad showing those faint snowy outlines in the far distance. It was not my imagination. It was a revelation. The planet was healing, and it didn't need us to teach it how. It only needed us to stop.

But humans are restless creatures. We cannot stay still for long. As soon as the world opened up again, so did our appetites: for travel, for markets, for noise. Within months, the same people who had once applauded doctors and delivery workers were back to honking at traffic lights, throwing garbage and chasing discounts. It's as if the global grief evaporated overnight. As soon as we could, we returned to choking the Earth. Even the spectacle of death could not humble us. We forgot how the whole world had shut down. We forgot the stillness that taught us humility. We forgot the empty skies, the silent airports, the sound of our own hearts. We forgot the fear that should have made us wiser.

The virus had touched every corner of the planet, every country, caste, creed and class. For once, humanity shared a single fate. But the lesson didn't last. Instead of becoming more conscious, we became more careless. Today, the air is thicker, the noise is louder and the waters are filthier. Even after staring death in the face, we went back to living like immortals.

While humans forgot, the trees remembered. They remember the months when the air was clean and the soil moist. They remember the quiet roads and the undisturbed nights when owls could hunt and bats could fly freely. They remember the hands that came daily with water cans while the rest of the world stayed hidden behind doors. Those were some of our best planting years. Not because we had more resources, but because we had fewer distractions. There were no ceremonies, no inaugurations, no banners. Just raw, unfiltered work. Our ground force teams moved like rivers: silent, steady and invisible. In those months, the trees taught us something profound: even in crisis, life goes on. When humans retreat, nature advances.

I often think back to that time, not with fear, but with a strange tenderness. The world had collectively slowed down, and in that slowing down was the seed of renewal. I remember thinking: *Maybe this is it. Maybe this is the turning point.* Maybe the pandemic would teach us the value of simplicity, of clean air, of fewer flights, of staying local, of growing food in our backyards. But our memory is short, our greed long. We broke every promise we made in those months. We flew more, mined more, consumed more. Governments that had once spoken of green recovery returned to coal. Cities that had breathed freely for the first time in decades are now choking again.

The lesson was lost not in years, but in months.

It's strange, isn't it? We watched pyres burning in the open, saw smoke rising not from factories but from cremation grounds. The air itself carried grief. Yet, as soon as the numbers went down, we turned away. We buried memory along with the dead.

The 'new normal' that everyone spoke of never came. Instead, we resurrected the old abnormal, the same patterns of pollution, waste and noise. Festivals returned, louder and more glittering than before. Roads are again full of plastic garlands and diesel fumes. We forgot how precious silence was. Maybe that's the curse of comfort: it breeds forgetfulness. When the danger passes, we erase the lessons, as if wisdom has an expiry date.

Chapter 40

Climate Change Is Here to Stay

Sit back and enjoy it! We are at a point where we may not be able to change what has already been done even if we really wanted to, and that sadly is a big if. The glaciers have received their memo, the oceans have taken offense and the atmosphere has decided it's done negotiating. We, the last species to arrive late to the party, are still holding climate conferences as if PowerPoint slides could persuade clouds. Enjoy the climate change now. What else can you do? Brace for it, as the airlines instruct before a crash landing: *head down, arms crossed and hope for the best.*

The most important thing now is to be physically fit. That's the new form of activism. Build your lungs, build your stamina, build your mental balance. Because when the heat waves turn sidewalks into griddles, you'll need both physical and philosophical fitness. When it rains for three days in October and snowflakes appear in the wrong hemisphere, you will need yoga, not panic. The body is your only home left with a manageable climate; keep it healthy.

If you can't control what happens outside, prepare from within. And then, do what you can in your personal radius. Green your home, your balcony, your terrace, your lane, your office corner, your factory courtyard, your local school, your college, your street. Don't get distracted by slogans about 'saving the planet'. The planet is not waiting for your rescue. It's been around for 4.5 billion years and has survived ice ages, super volcanoes and asteroid strikes. It's the humans who need saving. The planet is doing just fine; it's only rearranging the furniture to make room for the next tenants.

Sometimes, while sitting under a peepal tree, I feel nature has made up its mind. It has decided to do some culling. The biosphere, that grand orchestra of life, is simply tuning out the noisy instruments. Nature has been through extinctions before: five major ones, in fact. The Permian extinction wiped out 96 per cent of marine species. The Cretaceous one cleared the dinosaurs. We are just waiting for our own episode to be titled *The Anthropocene Extinction*, a name that sounds both scientific and poetic, like a museum plaque for the future.

Granny used to say, 'Nature has its ways.' She didn't read climate reports or IPCC summaries, yet she knew more than most experts. She would watch a river flood or a drought linger and whisper, '*Ab prakriti apna hisaab kar rahi hai*.' Nature is balancing its books. Even my parents, whenever I visit them in Pune, say it in simpler terms: 'Beta, nature has its own mind. It will do what is good for all species.'

We are just one species among twelve and a half million others. Nature cannot remain partial to us forever. We've been overdrawn in our ecological bank account, and now the system is sending reminders, rising seas, burning forests and disappearing birds. The account is approaching non-performing status.

Scientists have explained this with excruciating patience. Carbon dioxide levels are above 420 parts per million,[4] the highest in over 4 million years. Oceans are acidifying faster than any time in the last 55 million years.[5] Greenland's ice

---

[4] Krajick, Kevin, 'A New 66 Million-Year History of Carbon Dioxide Offers Little Comfort for Today', State of the Planet, 7 December 2023, https://news.climate.columbia.edu/2023/12/07/a-new-66-million-year-history-of-carbon-dioxide-offers-little-comfort-for-today/, accessed on 27 March 2026.

[5] 'Ocean Acidifying 10 Times Faster than Anytime in the Last 55 Million Years, Putting Polar Ecosystems at Risk', Mongabay Environmental News, 24 February 2014, https://news.mongabay.com/2014/02/ocean-acidifying-10-times-faster-than-anytime-in-the-last-55-million-years-putting-polar-ecosystems-at-risk/, accessed on 27 March 2026.

sheet is melting at 250 gigatons a year,[6] and the Himalayan glaciers are retreating at approximately 15m/year.[7] These are not numbers; these are nature's resignation letters.

Every IPCC report is like a polite obituary written for humanity, with footnotes. Yet, we respond by holding 'Climate Summits' in luxury hotels cooled by central air-conditioning and lit with LED chandeliers. Delegates arrive in private jets to discuss emission cuts. Look around. People are not panicking, they're planning trips. Climate change has created new genres of travel. Glacial tourism in Iceland. Coral-bleaching photography in the Great Barrier Reef. Even 'last chance safaris' in Africa, to see species before they vanish. A strange irony: We're not reversing climate change; we're romanticizing it. Apocalypse tourism is booming. Humans are the only species that can destroy their habitat and then book a guided tour to admire the ruins.

Social media calls it 'eco-conscious travel'. You offset your flight emissions by clicking a green button. Click, pay and fly guilt-free. Meanwhile, the Arctic melts quietly. Penguins look confused. Coral reefs turn white like ghosts of their former selves. The planet, in its stoic humour, seems to be saying, 'You go on posting reels, I'll go on rebooting.'

We are masters of denial. We build bigger air-conditioners to fight the heat they helped create. We plant lawns in deserts, pour concrete over fertile land and then

---

[6] 'NASA Scientific Visualization Studio | Greenland Ice Mass Loss 2002-2023', SVS, 7 July 2023, https://svs.gsfc.nasa.gov/31156/, accessed on 27 March 2026.

[7] 'Himalayan Glaciers Retreating at Varying Rates, Says Shri Kiren Rijiju', Pib.gov.in, 2023, https://www.pib.gov.in/PressReleaseIframePage.aspx?PRID=1941013&reg=3&lang=2, accessed on 27 March 2026.

wonder why groundwater has disappeared. We buy bottled water and complain about plastic waste. We hold 'Go Green' events where the banners are made of PVC flex sheets. Granny would have laughed. '*Beta, tumhe nature se zyada nautanki pasand hai*,' she'd say. (You people love theatrics more than the trees.)

And the politicians? They're still debating whether climate change is 'real'. Some believe it's a Chinese conspiracy. Others call it an economic distraction. Meanwhile, cities are drowning in floods one week and gasping in droughts the next. It's as if the weather itself has lost patience and decided to make every forecast wrong just to prove a point.

Nature, I believe, has already pressed the reset button. It has decided to let the temperature rise just enough to shake us off the branch. We were never the masters of this planet; we were tenants with unpaid rent. The forests we burned, the rivers we poisoned, the soil we exhausted—they are now reclaiming their space. COVID was just a trailer, a tiny reminder that one virus can pause the entire species. The full film is still in production.

Granny's prophecy echoes again: 'Nature cannot go on being partial toward humans.' Maybe it's time we accepted her wisdom. Maybe this chaos is not punishment but purification—a way for the planet to breathe again. Do what's possible. Plant a tree. Compost your waste. Cycle. Feed a bird. Grow a garden. Live lighter. But don't do it expecting applause or planetary gratitude. Do it because it's your way of saying thank you for the time you've had here. Your small green acts may not save the world, but they will save your soul.

And when you've done your bit, sit back under a tree, preferably a peepal, because it never judges, and enjoy the absurd beauty of it all. Watch the clouds drift through the polluted sky with poetic stubbornness. Watch the ants rebuild after every rain. Watch the weeds grow through cracks in the pavement. They are nature's quiet laughter. The planet will heal, with or without us.

Chapter 41

A Possibility Called Chandigarh

Chandigarh is the way a city should be. Yes, it is possible to have a city like Chandigarh. When the French architect Le Corbusier took to designing this magnificent city, our newly independent nation had barely found its footing. India had been partitioned; Punjab had lost its capital, Lahore. Pandit Nehru envisioned Chandigarh as the city of the future, a new symbol of modern India—clean, bold and full of light. 'Let this be a new town, unfettered by the traditions of the past,' he said. And Le Corbusier responded by creating more than a city; he created a philosophy of living.

He called it 'a city of the sun and of man'. It was not about architecture alone, but about equilibrium, between the built form and the breathing form. Chandigarh became a conversation between geometry and greenery, between human ambition and natural order. Each sector was a self-contained neighbourhood unit, with its own markets, schools and open parks. The green belts that separated sectors were like soft walls of foliage, living membranes that filtered the air and anchored the soil. The hierarchy of roads, from the V1S to the V7S, ensured smooth flow without chaos. It was a city designed to move like blood through veins: organized, quiet and alive.

The city's design principles were revolutionary: zoning regulations, open-space ratios, green buffers, walkable neighbourhoods and stormwater management—all woven with mathematical precision and aesthetic humility. The skyline bowed gently to the Shivalik Hills. The sunlight was invited, not fought. It was modernism with manners. And if such a city could rise seventy years ago, with slide rules and hand-drawn maps, why can't we build more today with satellites and supercomputers? The answer lies not in technology but in intent. We have the means, but we lack the will to dream.

Instead of creating new towns, we keep stuffing the old ones. Every year, millions migrate from rural India to urban centres, seeking work, dignity and survival. But our cities, already gasping, keep absorbing them without expanding their green or social infrastructure. We call this 'growth', but it's just overcrowding with better branding. We keep demolishing and rebuilding, wasting what already exists. Urban planning in India often looks like an endless loop of construction and correction, instead of foresight and preservation.

Australia offers an interesting mirror. Cities like Adelaide, Canberra and Melbourne are what Chandigarh would look like if it kept evolving. They have nature reserves every few blocks, sanctuaries for birds and small mammals. They treat open land as sacred breathing space, not as 'idle land'. Their land-use planning integrates ecology and economy. In India, unfortunately, we see every open patch as a potential parking lot or apartment tower.

We forget that cities are living beings. They need lungs, kidneys and heartlines, green cover, wetlands and mobility corridors. A healthy city is an ecosystem, not a concrete colony. Chandigarh was our first attempt at that ecosystem, a rare example of ecological urbanism before the term even existed.

I remember the old peepal tree near the parapet of Sukhna Lake. It is still there, ancient, vast and patient. Granny and I used to sit under it, her lap my first library. She would tell me about ghosts on ficus trees, and I would look up in awe at the branches that swayed and shimmered like spirits themselves. 'This tree has watched cities rise and fall,' she said once. 'And still it gives shade.'

She taught me early what most planners forget: a city is only as good as its trees. 'Birds, bees and butterflies,' she would whisper, 'are the real citizens of God. If they disappear, man will have no song left.'

Those stories were my first environmental lessons, long before I learned the word 'ecology'. Granny spoke of rivers, monsoons and soil as if they were her neighbours. 'Respect them,' she'd say, 'they know how to survive even when we forget.'

When I revisit Chandigarh today, I feel that same pulse. The geometry still breathes. The parks still hum. The urban canopy cover still cradles the roads. There's a discipline to the design that modern cities have lost. Chandigarh never shouts; it lives in quiet dialogue with its own air. And somewhere deep within me, that design became destiny.

When I moved later to Pune, it was Chandigarh that guided my hands. The discipline of sectors turned into the discipline of sapling rows. The green buffers became my roadside plantations. The geometry of streets translated into tree alignment techniques. My plantations on Range Hills Road were, unknowingly, my way of recreating Chandigarh's logic—order blended with life.

I planted where the glare was harsh, imagining shade the way Le Corbusier must have imagined sunlight. I drew invisible sectors in my mind, a playground of greens, each with its own rhythm of birds and bees. I realized then that fieldwork on two wheels was my continuation of that same vision: urban design done by hand, not by blueprint.

The red Speedking cycle I rode through dusty roads was my mobile planning tool. Where planners drew grids, I drew roots. Where governments discussed 'smart cities', I quietly grew small forests. Perhaps that's what Chandigarh really gave me—the belief that human settlements can coexist with forests. That you can live in a planned world without killing the spontaneous one. That modernity need not mean concrete. *Forest* means *for-rest*, Granny used to say. And I believe Chandigarh was India's first forest of ideas: a place built for rest, reflection and renewal. It is not just a city; it is an inspiration. A living lesson that structure need not suffocate life. That even amid geometry, trees can dance. And that a city born of vision can give birth to a lifelong mission: mine.

Chapter 42

If It's Green, It's Right

Principles of Happy Gardening

The mantra that has guided my life is simple enough to paint on a school wall, chant in a nursery or scribble on the back of a notebook: *Green hai toh sahi hai.* If it's green, it's right. I learnt that from the Earth itself. Every time I saw a patch of naked soil, cracked and gasping, I felt as though a living being had been stripped bare. I could almost hear the land's quiet plea for cover, for roots, for the soft weave of grass that protects its skin from burning. A naked land is a wounded land, and roots are the only healers nature has appointed. Nothing heals soil like the touch of roots. They are the fingers of the earth, stitching back its torn fabric.

When I first began to understand this, I didn't know I was echoing the wisdom of billions of years. Long before we arrived with our ploughs and plans, plants had already written their own constitution, wordless but binding. They decided that the Earth would not be owned, only shared; that sunlight was for all; that waste would be food again; that growth would come not from conquest, but from cooperation. It is a constitution we have violated at every step, and yet, miraculously, the plants still forgive us. They keep working, growing, mending, restoring, without judgment or pride.

My own education in this green law began with a flower. The shoe flower, they called it. A simple hibiscus that stood near the boundary wall of our school. It was my first red miracle. Its petals were the colour of sunset, soft and bold at once, and the golden stamens hung like tiny brushes dipped in light. We did not know its botanical name, nor did we care. Our class teacher told us that the hibiscus could polish black shoes, and that was all we needed to know.

We plucked the flowers, crushed the petals and rubbed them against our dull shoes until they gleamed. I remember the laughter, the smell of the petals, the squeak of shining shoes lined up outside the classroom. That was our first chemistry lesson, our first ritual of transformation. Nature, I realized, always offers herself for use, but never for misuse. She gives, but she expects respect.

Years later, when I was planting on roadsides and rescuing saplings from temple drains, the memory of that hibiscus would return. It taught me something vital—that beauty and utility are never separate. The same flower that attracts bees can also shine a shoe; the same leaf that breathes for the planet can feed a goat or shade a traveller. In nature, nothing exists for itself alone. Every stem, every seed, every root is in conversation with a thousand others.

It is a democracy so vast and decentralized that no parliament of humans can ever match it. Plants have no kings, no generals, no presidents. They thrive in a quiet anarchy of cooperation, each root aware of the others, each branch giving space for light. We, with our flags and borders, could learn a thing or two from that.

When I say '*Green hai toh sahi hai*,' I mean more than colour. I mean the right way of being. Green is a verb. It is the act of growing, of turning sunlight into life, of giving back more than you take. It is the opposite of greed. Every patch of green is a statement of balance. When grass grows over a wounded patch of land, it does not ask for applause. When a tree drops its leaves, it does not mourn the loss; it simply knows the cycle will continue. This quiet intelligence, this refusal to panic, is the true education we all need.

I have watched men and women exhaust themselves trying to stay on top of their work, as if life were a mountain and happiness waited only at the summit. They climb until their lungs give out, and when they finally stop, they realize that they have buried their joy somewhere along the way. The world tells us to rise, but the soil tells us to root.

One such person was an old lady I had known for two decades. A quiet woman with weary eyes and tired bones, she found her cure in a rooftop garden. Each morning, she climbed the narrow stairs with a watering can, and there, among pots of tulsi and money plants, she found her second wind. Her blood pressure stabilized, her anxiety softened and her laughter returned. She lived another decade, dying one afternoon among her plants, sitting on her stool, a leaf in her lap. The doctor said her heart gave way. I think it simply went back to the soil.

Plants are generous landlords. They let us breathe their waste and eat their bodies. Whether we live on vegetables or meat, everything begins with plants. The animal that feeds us eats the grass that feeds the world. We are all made of green dust rearranged into human form. Once you understand that, you begin to see yourself differently. You stop standing apart from nature and start standing within it. You realize that your own body is a rented structure built from borrowed elements, carbon from trees, calcium from shells, iron from soil. When you understand that, humility follows naturally.

The human race has forgotten this debt. We build without asking, consume without returning and label our thefts as 'progress'. But the law of the green world is simple: whatever cannot be regenerated must not be taken. The

Earth does not forgive permanent damage. Soil stripped of its microbiome will take centuries to recover. A cut forest does not grow back as the same forest.

What we call development is often just delayed decay. To live rightly, we must learn from our roots: how they take only what they need, how they give back stability, how they never stop communicating underground. A forest is not a collection of trees; it is a community of roots holding hands.

I once heard someone say that traffic rules are like ecological laws, learned only after a few small accidents. Perhaps that's why it's good if young people face minor scrapes early in life, so they learn to steer carefully. Life, too, teaches through collisions. We bruise, we adjust, we develop reflexes. The ecosystem is no different. When we clear too much land, flood too many valleys, poison too many rivers, the Earth reacts. Floods, droughts, diseases—they are its ways of braking suddenly, warning us to drive better.

Back in 2003, when we started visiting schools across Delhi with a simple message, *grow something*, we didn't realize that we were beginning a movement. We weren't invited by any ministry. We simply walked into government schools and said, 'We'd like to talk to your children about growing something.' The principals looked at us with surprise and then smiled. They said, 'Please do.' The response was overwhelming. We carried saplings, seeds and small pots. We told stories, asked questions and watched their eyes light up. It was like opening Pandora's Box, except that what flew out were butterflies, not curses.

From one school to another, the campaign spread like pollen in the wind. The message was simple enough for a five-year-old to remember and profound enough for a teacher to quote in an assembly. *Grow something.* Grow a plant, a habit, a sense of wonder. Grow respect for the green nation we are all citizens of.

In those classrooms, I saw the future shift slightly toward sanity. Children who had never touched soil before dug their fingers into it. Girls who were afraid of worms began feeding them. Boys who laughed at plants started naming them. When a sapling survived a harsh summer, they clapped. When it wilted, they mourned. These were small awakenings, but they mattered.

A school garden is not just a project; it is a lesson in the truth of life itself. Every child who plants a seed learns the first article of the green constitution: the Earth belongs equally to all living beings.

This constitution does not have the Ten Commandments or the seven deadly sins. Its principles are simple, almost self-evident: respect diversity, avoid waste, share resources, regenerate what you take and never break the chain of life. The plants have been living by these rules since time began, while we, their younger cousins, keep writing new laws and forgetting the old ones. If humans are to survive the coming centuries, we will have to naturalize ourselves as citizens of the plant world again.

Every act of planting is a vote cast in that direction. When you place a sapling in the ground, you are participating in the oldest democracy on Earth. You are saying, 'I belong.' The soil does not care for your religion or your passport. It only cares whether you can nurture it. The air that leaves your lungs becomes the air a tree inhales. The circle closes beautifully, without bureaucracy. Perhaps this is why gardening is so deeply satisfying: it restores a sense of meaning that paperwork and ambition cannot.

Sometimes, standing by a roadside, watching the wind ripple through a patch of grass I once planted, I feel like I am reading a text written in light and movement. Each blade bends, but the field holds steady. This, I realize, is resilience. Not the loud defiance of monuments, but the quiet persistence of green.

A forest doesn't protest; it simply returns. A plant cut today sprouts again tomorrow, not from vengeance but from necessity. That is the spirit of life. If we could live with that spirit, giving, growing and forgiving, the planet would heal faster than we think.

The day humanity learns to think like plants, there will be no wars over borders, no hunger born of greed, no loneliness in the cities. Plants have mastered coexistence. They share space, light and water without envy. They understand something we have forgotten: that strength lies not in standing above others, but alongside them. The highest tree still depends on the smallest root. The humblest moss holds moisture for a mountain. Everything connects.

Even the act of rest in nature has dignity. When trees shed their leaves, they are not dying; they are making room. When rivers flood, they are not angry; they are redistributing. When soil cracks, it is calling for attention, not punishment. The universe runs on dialogue, not dominance. The more we listen, the more we understand that there is no hierarchy of existence. A peepal leaf, a pigeon, a pebble and a person—all are instruments in the same orchestra.

Sometimes I wonder if the world would be calmer if every human had to plant at least one tree before being allowed to speak in public. Perhaps governance would improve if politicians spent one hour each day watering a sapling. Perhaps education would make more sense if every student learnt to read the alphabet of leaves before writing essays about GDP.

We've built skyscrapers tall enough to touch clouds, but forgotten how to kneel and touch the soil. The wisdom of the green world lies in its posture: never arrogant, always grounded. The plants, for all their silence, are the most eloquent teachers. They show us that growth need not be noisy, that strength can be soft, that survival is not a competition but a conversation. The Earth's oldest citizens have already solved the problems we are still debating—sustainability, equality, peace and renewal.

When I look back on my decades of planting, I see not just trees but lessons standing upright. Every peepal I have planted has taught me patience. Every neem has shown me endurance. Every patch of grass whispered, 'Do your work and step aside.' That is the real mantra behind *Green hai toh sahi hai*. Do what needs to be done; the results will take care of themselves.

We are temporary gardeners in an eternal garden. Our duty is not to dominate but to tend. Soil must never be left naked. Roots must always be allowed to breathe. And if, some day, humanity disappears, the plants will not hold grudges. They will simply cover our ruins with green, as they have always done, making things right again in their quiet way.

So, I plant. Because planting is prayer. Because roots are the original healers. Because the Earth deserves its cover, and we deserve the peace that comes from giving it. *Green hai toh theek hai*. Always has been. Always will be.

Chapter 43

Afforestation

Afforestation is not a government project or a modern buzzword. It is an act of remembering. To plant a tree is to remember what the soil still knows: that life can begin again from dust. In my lifetime of planting millions of trees across India's highways, villages and cities, I have seen forests being born in many ways. Some rise slowly, as if shy to reveal themselves. Others rush up in excitement, eager to fill the void humans have created. Each forest method, old or new, traditional or Miyawaki, is an interpretation of that same memory; a different verse of the same hymn sung to the Earth.

When I began planting in the 1970s, we knew only one way: the traditional way. It was slow, patient and deeply personal. You dug a pit, sometimes with a borrowed shovel or even your hands. You filled it with loose soil, added compost, planted the sapling and protected it with stones. And then you waited. Every monsoon was a festival, every survival a small victory.

Traditional afforestation has its own rhythm, like classical music, steady and layered. The roots go deep, the branches take years to stretch, and yet the harmony lasts for generations. You could plant a neem or a jamun in a dusty village and return after twenty years to find a shade that shelters farmers, vendors and wanderers. It teaches you patience, gratitude and humility.

The disadvantages are only of our own making—impatience, neglect or lack of care. These plantations take time. They need water, fencing and protection. They are vulnerable to grazing animals and careless hands. But once they survive the first few years, they become immortal.

Then came the Miyawaki forests: fast-growing, dense and full of promise. When I first saw one, I was intrigued. It looked like a pocket-sized jungle, like nature squeezed into a corner of concrete. Developed by Japanese botanist Akira Miyawaki, this method plants dozens of native species closely together to recreate a natural forest environment. The soil is enriched, the saplings are densely packed and within three years, you get a mini-forest where birds return, insects buzz and the air feels alive again. I have seen Miyawaki forests rise on school campuses, corporate premises and even roadside plots that were once garbage dumps.

For a country gasping for oxygen, Miyawaki forests are like green lungs inserted into the city's chest. Their biggest advantage is speed. What takes a traditional forest a decade to achieve, Miyawaki can do in three years. It is ideal for cities where space is scarce and patience scarcer. It works beautifully when you want to create biodiversity in limited areas. But it comes at a price—a high cost of soil preparation, continuous watering and professional supervision.

It is not for every region. In dry belts, the dense plantation suffers, the roots compete and some species die early. Yet, it is a revolution worth embracing in urban areas. I always say, if you cannot grow a jungle, at least grow a memory of it.

In rural India, the forest wears another form: agroforestry. It is the farmer's forest, where trees and crops coexist. Here, the neem guards the field boundary, the amla gives fruit and the moringa provides leaves for food and medicine. I have always admired the wisdom of rural people who understood that trees are not enemies of agriculture. They are companions. Their shade keeps the soil moist, their roots prevent erosion and their fallen leaves become natural manure. I remember working with farmers in Bundelkhand and Haryana who were hesitant to plant trees, afraid they would take away space or nutrients. But once they saw the results—healthier soil, better moisture retention and even improved yields—they became lifelong advocates. Agroforestry is the perfect example of ecology and economy working hand in hand. It is slow afforestation, but it sustains both soil and soul.

In cities, afforestation is no longer a luxury. It is survival. Urban afforestation, whether through avenue plantations, rooftop nurseries or Miyawaki pockets, is the only way to make our cities liveable again. I have seen the magic of a few trees transforming entire neighbourhoods. A row of peepals along a school boundary can drop the temperature by two degrees. A handful of neem trees along a dusty road can filter the air better than any machine. Every patch of green matters.

Urban afforestation faces its own challenges: lack of space, pollution and short-term thinking. But the results are immediate and visible. The soil breathes again, butterflies return and even humans begin to walk more slowly. Cities need not be barren; they just need to share space. I often tell urban planners to give nature half the city, and it will take care of the other half.

Then there are the forests planted for water. Along rivers, lakes and slopes, trees become guardians of flow. Planting near rivers is like writing poetry for the unseen.

You may never meet the beneficiaries, but you know the trees are holding the soil, filtering the rain and recharging the aquifers below. I have planted along the Yamuna, Ganga and smaller rivers whose names have faded from maps. In these zones, the forest is not a showpiece; it is a necessity. It prevents floods, stabilizes slopes and keeps the land alive. The roots of a peepal can hold a hillside better than a wall of cement.

And, of course, there is the controversial child of modern forestry: compensatory afforestation. This is what the forest industry pays for when it cuts another. It sounds fair: you destroy one forest, you plant another. But nature doesn't balance equations that way. A forest is not a number; it is an organism, a civilization of its own. You can't replace a 200-year-old banyan ecosystem with a row of eucalyptus on a barren plain. Yet, this system has forced corporations and bureaucrats to think green, even if mechanically at first. Over time, many projects have evolved into genuine ecological models, especially when local communities are involved.

The key is diversity, not planting just one species, but recreating the mixed forest tapestry.

Afforestation is not a one-size-fits-all mission. The traditional forest teaches patience, the Miyawaki forest teaches speed, the farmer's forest teaches balance, the urban forest teaches adaptation and the river forest teaches humility. Together, they form a symphony. Each has its strengths and flaws, but collectively they remind us that no single method can heal the planet. Healing requires diversity, just like the forest itself.

I often think of my work as a form of storytelling: stories told through roots and leaves instead of words. When I look at a stretch of trees planted years ago, I see chapters of effort, seasons of struggle and pages written in soil. The peepal grove along the Ring Road, the neem belt in Haryana, the Miyawaki forest in Gurgaon, the amla orchard in Bundelkhand are all verses of one long poem that I have been writing since I was a boy with a sapling in his hand.

Each time I plant, I learn anew that forests are not born in nurseries. They are born with intent. You can train people, design models and analyse carbon data, but unless the planter feels love, the forest will never truly grow. Machines can dig, and drones can drop seeds, but only the human heart can create belonging. In the end, it doesn't matter which method you choose: traditional, Miyawaki, agroforestry or urban pockets. What matters is that you plant. The Earth doesn't care about our techniques; it cares about our tenderness. Every act of planting is an act of faith. A sapling planted with love will always find a way to live, no matter how harsh the climate or how limited the space. Planting is our way of saying to the future: *we tried to make it liveable for you.*

If you have one acre, plant a forest the old way. If you have one square metre, make it Miyawaki. If you have a field, mix trees with crops. If you have a balcony, grow a small nursery. And if you have nothing but time, give it to a tree. The planet doesn't need our opinions, it needs our hands.

Afforestation, in all its forms, is humanity's most beautiful apology. We have taken too much. The least we can do now is give shade in return.

Chapter 44

Urban Greens

If tree planting were a sport, I suppose I've been in the field long enough to earn a few disqualifications. I've buried more mistakes than saplings, and that's saying something. But that's how afforestation science really grows—not in labs, but in patches of failed experiments that slowly turn into fertile soil.

Most people believe cities are where forests come to die. I have learned the opposite: that cities are where forests come to reinvent themselves. Delhi, for example, is not short of space; it is short of imagination. The pavements, dividers, roundabouts, school yards, parks, rooftops—they're all unclaimed green laboratories waiting for intent.

When we started *Give Me Trees* Trust, the science was basic and the faith was excessive. We didn't have GIS mapping, hydrological models or carbon calculators. What we had were hands, buckets and a deep suspicion that the Earth would help if we helped first. Over time, patterns emerged. A roadside sapling's survival rate depended not on rainfall but on routine—watering twice a week, loosening the soil every ten days and replacing the mulch every month. We began to treat tree survival as a dataset, not a miracle. The results were startling. With consistent aftercare, survival rose from 45 per cent to nearly 85 per cent. The conclusion was embarrassingly simple: trees don't die of drought; they die of neglect.

Urban microclimates are temperamental. One street corner bakes under reflected glass heat; another stays cool because of an old neem. I began to see the city as a patchwork of thermal zones. The science of planting had to adapt.

Afforestation is not about replicating; it is about reading. On Delhi's Ring Road, the soil was too sandy, so we added clay. Near Gurgaon, it was heavy with construction debris, so we added compost and sand to breathe. In floodplains, we elevated the pits by 30 centimetres to prevent waterlogging. Each site has a pulse; you just have to listen.

During one planting drive in Mayur Vihar, we noticed that peepal saplings grew faster near bus stops. The reason, I joked, was spiritual; they loved gossip. The scientific explanation was carbon dioxide concentration. Commuters exhaling near the trees created a micro-CO_2 enrichment zone. The trees thanked us in silence by growing faster.

We tested everything—linear, cluster, Miyawaki and mixed-forest models. The Miyawaki model, dense and ambitious, worked beautifully in controlled plots but failed in open city areas where water scarcity and root competition made maintenance impossible. The mixed-forest model, however, proved sustainable. By pairing fast growers (like subabul and cassia) with slow elders (like peepal and neem), we allowed shade succession to occur naturally.

On paper, this looked like forest science. In practise, it felt like matchmaking: who gets along with whom under the same patch of sky. The neem doesn't like overwatering, the peepal loves sunlight, the jamun prefers companionship and the gulmohar just wants attention.

We discovered that biodiversity was not just ecological insurance; it was emotional insurance too. Volunteers got attached to species differently. Someone would guard a gulmohar like a family heirloom, another would speak to a peepal as if it were a sibling. When emotion enters data, sustainability begins.

Delhi's summers taught us the true meaning of scarcity. At 46°C, no amount of optimism can replace hydration. So, we began experimenting with micro-irrigation models, clay pot watering, wick irrigation and bottle-drip systems. We buried terracotta pots near saplings and filled them weekly; the slow seepage kept the soil moist for days. It was an old village trick with a modern name: *subsurface irrigation*. Later, we adapted the same method for rooftop gardens using cotton wicks dipped in water-filled bottles. It was low-tech, high-return ecology. The city needs such science—small, local, patient innovations—that don't depend on billion-dollar budgets. Afforestation will never be solved by drones; it will be solved by discipline.

From the soil to the sky, the journey felt natural. My barsati rooftop became a prototype for what I began calling 'Sky Forests'. Every plant up there was both student and teacher. We measured temperature drops of 2–3°C after adding forty pots of mixed vegetation. The humidity increased slightly, bird visits multiplied and the water runoff during rains reduced by nearly half.

It's easy to talk about global warming in air-conditioned halls. It's harder, but far more satisfying, to cool your own rooftop. Urban forestry, I realized, is not just about trees; it's about microclimate modification. Every leaf is a miniature air conditioner. Every rooftop that grows green contributes to what scientists call 'urban heat island mitigation'. I prefer to call it 'rooftop redemption'.

What the textbooks never tell you is that roots have memory. You transplant them roughly, and they sulk for weeks. You plant them with calm hands, and they settle instantly. I've watched roots trace cracks in old pavements, find forgotten moisture in cement crevices and cling to life where logic ends. Science can explain capillary action and osmosis, but not the stubbornness of a banyan root looking for hope. In the end, afforestation is not a science of planting; it's a science of persistence. The first year is about protection, the second about adjustment, the third about independence. After that, the tree teaches you.

Our volunteers often ask me, 'Sir, when will this become a forest?' I tell them, 'When you stop counting trees and start counting shade.'

We once tried planting along the Yamuna floodplains. The river had other plans. During the first monsoon, half our saplings were washed away. In the second year, we raised them on mounds. The third year, they held. Today, those very mounds are green islands of hope.

The field data from such sites often looks unimpressive, with survival rates between 70 and 80 per cent. But numbers are deceptive. Each surviving tree creates a micro-habitat for insects, birds, fungi and grass—an exponential ecological multiplier. A single mature peepal supports more life than a spreadsheet can count.

Science may measure photosynthesis in micromoles and transpiration in grams per square metre, but on the ground, it smells of sweat, mud and burnt rubber. I've spent a lifetime trying to make these two worlds shake hands: the academic and the earthy, the measured and the messy.

Urban afforestation will not come from policies alone; it will come from rooftops, medians and sidewalks, from people who remember that the soil beneath the city is not dead, but only waiting.

Chapter 45

Nature As Co-Governor of Cities

Justice must be given to the green areas. That has been my lifelong dream, a dream of equality between the built and the living, between human and natural. A world that is fifty-fifty. Fifty per cent governed by the natural world and 50 per cent by humans. That much sharing we must learn in our urban areas, in our cities and in our daily lives. Only then can we sustain. Only then can we survive.

It sounds simple: half for us, half for the planet. But in that simplicity lies the deepest truth of existence. The Earth has always maintained its balance between land and water, day and night, light and shadow, predator and prey. Only human beings broke that code of equality. We took more than our share: more water, more soil, more air, until even the wind began to feel tired.

Fifty-fifty is not an emotional number; it is a biological necessity. Every living system needs balance to function. The moment one side dominates, collapse begins. We have built cities where cement breathes more than soil, where traffic outnumbers trees and where every open ground becomes a potential construction site. We call it development, but what we are really doing is upsetting the balance, brick by brick.

When I walk through modern cities, I often feel as if the Earth is holding its breath. The roads radiate heat, the drains choke and the sky seems to shrink behind glass towers. We speak of progress, but we have forgotten that survival itself is the highest form of progress. Look closely at any major city today. Delhi, Mumbai, Bengaluru, Chennai—each one tells the same story in different accents: the story of forgotten rivers, vanished ponds and stolen greens.

The fifty-fifty balance has been broken so badly that most cities now have less than 15 per cent green cover. A city that should have had half its space breathing freely has sealed itself in asphalt. That imbalance shows up in our health, rising heat, polluted air, floods in the monsoon, drought in summer and an epidemic of anxiety and breathlessness. Heat islands form wherever trees vanish. A patch of bare concrete can be five to ten degrees hotter than a tree-covered zone. Multiply that by millions of square metres, and you understand why cities now feel like ovens. The tragedy is that we have normalized this suffocation.

The signs are all around us—sparrows gone missing, frogs falling silent, drains flooding with the first rain and children growing up with inhalers instead of slingshots. This is what happens when the fifty-fifty law of nature is ignored. Cities forget to breathe, and so do their citizens.

Even economics echoes the same truth. The most expensive localities in any city are those closest to parks, lakes or tree-lined boulevards. The logic is simple: people pay for peace, oxygen, silence and beauty. Yet we still treat greenery as a decorative luxury, not the foundation of value. I often say: The only thing real about real estate is greenery. Everything else is imitation—the concrete, the marble, the metal, the branding, the brochures. All of it can crumble, all of it can fade. But the value of a tree, once planted, only increases with time. A peepal planted today will be a monument of air and shade a century later.

Developers spend fortunes on artificial landscapes, vertical gardens and rooftop lawns, yet they forget that the real wealth is in soil that is alive. The rest is an illusion built on borrowed time. Because what nature creates is the only authentic part of what we call property.

Nature must not remain a visitor in the house of governance; she must be a co-governor. Fifty per cent of every city must be governed by the natural world. What if every urban masterplan had an ecological seat, a ministry of trees and water bodies with legal authority over half the land? What if every new project had to pass a balance audit, proving that its green footprint matched its built footprint? Imagine if every township or housing society were designed like an ecosystem: half built, half breathing. Roads lined with avenue trees, every block opening into a community garden, every rooftop growing food and every rainwater drain leading into a restored pond. That would be a real smart city; not one ruled by gadgets and sensors, but by shade and silence.

The world keeps chasing new technologies, but no machine has yet equalled the efficiency of a tree. Nature is the original engineer. We need only give her room to work.

This principle of balance is not a dreamer's poetry; it is scientific design. Having half the city under natural cover ensures climate stability. Trees regulate temperature through transpiration. Water bodies recharge aquifers. Open soil allows percolation and prevents flooding. Biodiversity corridors maintain insect populations that pollinate our crops. Even mental health studies show that access to greenery reduces anxiety, improves concentration and lowers crime.

Urban planners often treat parks as amenities, but they are in fact public health infrastructure. The oxygen from trees reduces particulate matter. The moisture from lakes cools the air. The microbes in healthy soil regulate air chemistry. Every living patch of Earth is performing silent, unpaid labour for the city. If we gave back 50 per cent of space to the natural world, she would repay us a thousandfold in air, in water, in sanity.

Chandigarh was built with the idea of providing breathing spaces. Le Corbusier's grid included gardens, lakes and tree-lined avenues. The city still feels calmer because it respects open ground as sacred. But even Chandigarh is under pressure now, with new suburbs eating into its peripheries. Bengaluru, once the city of lakes, thrived on its water network. Every lake overflowed into another, creating a self-cleaning system. When builders filled them up, floods followed. It was not punishment from heaven; it was simple hydrology.

Singapore offers another model. It calls itself not a 'garden city' but a 'city in a garden'. That subtle shift of words changes everything. Every road, roof and wall is seen as potential green cover. That is fifty-fifty justice in practise. I have seen glimpses of this balance even in our own work at *Give Me Trees* Trust. From Mayur Vihar's rooftops to Karma Lakelands' greens, from small village nurseries to city school plantations, each patch of reclaimed nature becomes a heartbeat of balance. The idea is the same: don't fight for space, share it. The idea of fifty-fifty justice is not only scientific or economic; it is deeply spiritual.

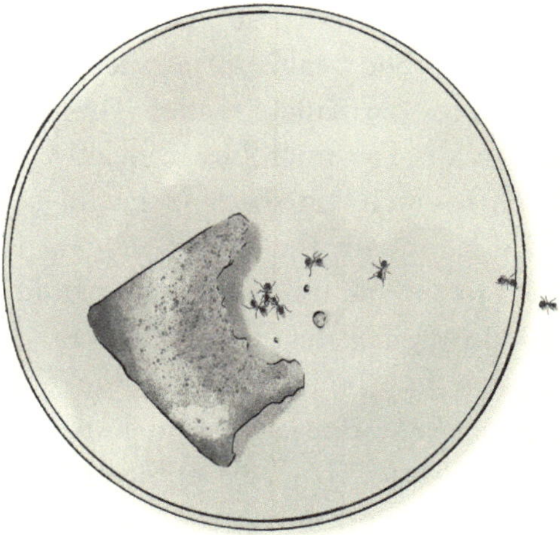

Granny used to say, 'Never eat everything on your plate; leave a little for the ants.' That was her version of fifty-fifty. She believed every act of sharing kept the world in balance. The same applies to land and sky. Leave half for others, for birds, for soil, for rivers. In our scriptures, too, there is an echo of this law. The Upanishads speak of balance between the seen and the unseen worlds, between desire and duty. Buddha found enlightenment under a tree, not in a temple. The tree became the teacher because it embodied equilibrium: rooted in earth, reaching for the sky, taking only what it needs, giving everything back.

We do not need new philosophies. We only need to remember the old ones. I dream of cities where schools are shaded by peepal and neem, where buses stop under trees, where offices have gardens instead of smoking zones, where ponds are community centres and where birds are not visitors but residents. Let our cities become forests of cohabitation. Every citizen should be able to walk to a green space within five minutes. Every rooftop should hold a pot or two. Every government office should adopt a grove. Every corporate building should plant its conscience along with its logo. This is not an impossible vision. It only demands that we change the definition of development. Let us measure growth not by square feet of cement but by square feet of shade. Let GDP also mean Green Domestic Product.

When 50 per cent of urban space is governed by nature, the other fifty will automatically flourish. The city will breathe again, not as a machine, but as a living organism. We can begin today, not by waiting for policies, but by planting and protecting. One tree, one pond, one rooftop at a time.

We often speak of saving nature, but perhaps it is she who keeps saving us, silently, again and again. When the air clears after rain, when the first leaf sprouts through a crack in the pavement, when a bird builds its nest on an electric wire, those are nature's small gestures of forgiveness. She does not hold grudges. She only asks for space. Fifty-fifty space is all she needs. Half for her, half for us. In that sharing lies the justice of survival. There is no philosophy deeper than this. It is obvious. It is urgent. Nature is the only saviour.

Chapter 46

Education Is the Key

We are drowning in waste not because we lack technology but because we lack tenderness. Machines can process garbage, but only a sensitive human being can prevent it. Sensitivity is the best technology we will ever have. Real revolution does not come from laboratories or ministries. It comes from people: from citizens who decide to live consciously. Citizen science is not about collecting data; it is about collecting awareness. It is when a farmer notices a new bird species in his field, when a schoolchild records the rainfall in her village, when a homemaker composts kitchen waste instead of throwing it into a plastic bag—that is the beginning of citizen science. It is the science of participation. When people understand that the planet is not an abstraction but an extension of themselves, everything changes. That shift cannot be forced. It must be taught, nurtured and celebrated. It must begin early: in schools, in homes and in stories.

As a child, I remember watching my grandmother compost fallen leaves. Over weeks, they would shrink, darken and become rich soil. She used to say, '*Kuch bhi bekar nahi hota, bas waqt chahiye* (Nothing is useless; it only needs time).' That was my first lesson in waste management: not from a textbook, but from an old woman with a broom and deep wisdom. Today we call it 'organic recycling' or 'bio-composting', but back then it was just common sense. My grandmother didn't use the term 'zero waste', but she lived it.

In all my travels across twenty-one states and 226 districts, I've met two kinds of people: those who believe blindly and those who observe patiently. The real transformation happens when belief and observation meet. To be spiritual is to be sensitive; to be scientific is to be sensible. Together, they form the true intelligence of a human being. The scientific temper that I speak of is not about laboratories and white coats. It is about the courage to question, observe, record, experiment and, above all, care.

If we can raise a generation that observes the way a bird drinks water, that records the flowering of a peepal tree, that understands why soil smells the way it does after rain, we would not need to preach about climate change. They would already be living the solution.

Education: The First Tree We Must Plant

If I had the power to direct the destiny of a nation, I would invest everything, not in missiles or skyscrapers or sports leagues, but in elementary education. Elementary education is the root system of all learning. Higher and specialized studies come much later. By then, we have already formed our habits, our curiosities, our fears and our sense of wonder. Whatever happens, happens before the eighth grade. That is the soil in which the mind first takes shape.

I say this out of my own life experience. I lived through it. I was born out of it. If my teacher, Mrs Williams, had not been around us in those early years, my destiny would have been entirely different. She taught us not only to read and write, but to *observe*, to notice a caterpillar on a leaf, the sound of rain on the roof, the smell of damp soil after a shower. Those were not mere lessons; they were awakenings.

India's greatest strength has always been its teachers—from the quiet guru under a banyan tree to the humble schoolteacher in a rural classroom. If we nurture them, everything else will follow because everything begins there. Primary school teachers are like Gods, creating new citizens for the country. They build young minds, shape attitudes, not just alphabets. They make the first gardeners of the mind. They do not merely plant information; they plant the seeds of sensitivity and compassion, the spark of curiosity, and the rhythm of learning.

We underestimate the power of a teacher with a piece of chalk. We overlook the miracle of a child's first question. Those are the moments when civilizations are built. Those moments become the real roots that hold a civilization together. If we water those roots well, the nation will grow on its own. A teacher is a tree—patient, rooted, giving shade and fruit without asking who sits beneath. A student is the soil—eager, porous, ready to receive.

We underestimate what happens when a teacher explains photosynthesis to a group of ten-year-olds, and they look at a leaf with wonder. But, in reality, that is the moment when climate awareness truly begins. When a child learns to use both sides of paper, not because someone forced them but because they learnt to value trees. That is real education.

If we can bring such teachers and students together in every village and every city, the future will take care of itself. A nation that worships its teachers and respects its students will never go hungry or poor. Invest in early education, and you will never have to worry about the economy, defence or politics. Those are natural outcomes of an educated, aware population.

Make gardening a compulsory subject in every elementary school, not as a hobby class once a year, but as a way of life woven into the timetable. Children should know the patience of a seed, the humility of soil and the quiet miracle of growth. They should learn how sunlight feeds, how worms work unseen, how weeds compete and how failure is as natural as flowering. In a garden, there are no exams, but every plant is a test of attention. There are no ranks, but the careless hand is soon corrected by the withering leaf. Gardening teaches what no textbook can: the relationship between effort and time.

If a child learns to care for a plant, that same child will grow up caring for people, rivers, forests and nations. The habit of nurturing is not something you acquire in adulthood; it begins with holding a seed in your small palm and watching it sprout. That is why I often say: give me a spade instead of a syllabus, a patch of soil instead of a paragraph, and I will show you a classroom more alive than any with four walls. When I look at my own life, planting trees, teaching volunteers, writing on scrap paper, I see it as one continuous act of education. I am still that student under the open sky, learning from soil, leaves and light.

Throughout my life, I have saved every scrap of paper that carried my thoughts: thousands of diaries, notebooks and loose sheets. Some were clean, some stained with tea or soil. Many were written in the field, sitting under a tree or beside a nursery bed. There were days when I wrote with muddy fingers, with a pencil borrowed from a worker or a volunteer. There were days when I had no notebook at all, and so the back of an envelope or the edge of a newspaper became my page. Those fragments are now my history. When I look at them now, I realize how close ecology and economy are; both depend on how we use and reuse. A piece of paper is a small economy of thought; it records what might otherwise be lost. Even now, in the age of phones and screens, I find myself drawn to old paper. There is something humbling about it: its weight, its silence, its readiness to receive. When I write on reused paper, I feel I am not creating waste; I am continuing a conversation.

Every session I conduct is rooted in experience, sometimes scientific, sometimes poetic. I show them how to compost, how to plant, how to observe. But more importantly, I show them how to *feel* what I feel, to appreciate how even nature writes on both sides. The fallen leaf becomes mulch; the decayed fruit becomes soil. Nothing is one-sided in the natural world. Everything has a reverse, a reflection, a return.

We speak of sustainability as if it were a new word. But for me, sustainability is just remembering how to live simply. How to reuse a clip, how to replant a sapling, how to reimagine an old habit. When I hold up a small piece of paper during a session and say, 'This is not trash, this is a tree,' I see their eyes widen. At that moment, something shifts inside them.

Large problems, like sustainability, often frighten us because we imagine their solutions must also be large. But I have learned that all revolutions begin with a small act—reusing a sheet, planting one sapling or refusing one plastic bag. Each act may seem insignificant, but together they form a culture. The culture of care. The culture of consciousness. The culture of continuation. When you write on the back of an envelope, you are quietly declaring: *I still believe in the other side.* When you plant a tree on a barren roadside, you are saying: *I still believe in tomorrow.* And when you teach a child to observe an ant, to protect a leaf, to feel the pulse of the Earth, you are saying: *I still believe in education.*

Sometimes I look at all my notes—scattered, uneven, written in different years on different scraps—and I wonder what they would look like if put together. Perhaps they would form a forest of thoughts: uneven, wild and alive. Each piece of paper is like a sapling of memory. Together, they make sense.

The world today is obsessed with perfection and uniformity. But nature thrives on imperfection and diversity. A reused page, a rough edge, a faded ink line—these are not flaws, they are signs of life. Our future depends not on perfection but on participation.

Chapter 47

A Time for Green Elections

Green Parties and Green Ballot. Why not?

The time has come to vote for the colour of life—green. Not the pale green of cosmetic promises or recycled slogans, but the deep, chlorophyll-rich green of sincerity. The green that photosynthesizes hope, cleans the air and outlasts every speech. For too long, elections have been festivals of noise and plastic. Flags of every colour but none of the leaves. Candidates distribute pamphlets that later choke drains, while promising to make the city 'smart'. The forests listen in silence. The rivers roll their eyes.

We need leaders who can talk about the oxygen budget as easily as they talk about the fiscal one. Who understand that economy and ecology are not enemies, but siblings born of the same syllable—*eco*—meaning home. Those who can say, without a teleprompter, that a polluted constituency is a defeated one.

In Germany, when the Grüne Partei (Green Party) first entered Parliament in 1983, they walked in wearing jeans and carrying flowers. The establishment laughed. The same establishment now reads their environmental reports before drafting budgets. In Australia, the Greens managed to push climate change into the heart of political debate, not through protests alone, but through persistence. In New Zealand, they declared rivers as legal persons, a concept more humane than most human laws. In Latin America, Evo Morales of Bolivia spoke of 'Pachamama', Mother Earth, in the Constitution itself. In Bhutan, happiness became a measurable index only when forests were protected.

And what about India, the land that first called rivers 'Maa' and trees 'Devi'? We, who light lamps under peepal trees, now cut them for parking lots. We, who once worshipped rainclouds, now curse them for traffic jams. Perhaps it is time we voted like we worship: with reverence, not rhetoric.

We need representatives who talk of greens, renewable energy sources and the welfare of the community. Who speak not only for human rights but also for the rights of rivers, forests and other species. Who can stand up and say, without embarrassment, that protecting wetlands is more patriotic than building statues. We need young politicians who carry tree saplings, not cricket bats, in their campaigns. Who promise to protect not just votes but roots. Who talk for the environment, ecology and pollution control and who understand that *the environment is not outside us, it is within us!*

When lungs cough, it is not Delhi's problem; it is democracy's asthma. When groundwater drops, it is not Haryana's problem; it is India's thirst. When the monsoon fails, it is not the farmer's fate; it is our collective failure of faith.

Imagine an election rally where candidates arrive in bullock carts, not SUVs, to save diesel. Imagine manifestos printed on recycled paper with seeds embedded in them. After reading, you plant the promise. Imagine parliamentary debates that begin with, 'Before we start, let's check the air quality of this hall.' Imagine if a politician planted a tree every time they broke a promise, we would have forests overnight!

We need to make ecology sexy again, not just as a cause, but as a culture. Young leaders who see composting as cool, rooftop solar panels as status symbols and water harvesting as the new yoga.

The Greens of Finland campaigned for 'Rights of Future Generations'. The Scottish Greens pushed for a ban on single-use plastics before many big economies did. The Taiwan Green Party runs schools for environmental literacy. And the African Green Network has reforested more than 10,000 acres of savannah through community-led actions. All this without billion-dollar campaigns, but with conviction. Because the truth is: *Green is not a party colour; it is the colour of survival.*

India deserves a new kind of politics: one that grows out of compost, not corruption. One that speaks of solar panels on school rooftops, afforestation drives as employment programmes, waste management as women's empowerment and green cover as a national security measure. We need leaders who can merge the Ministry of Environment with the Ministry of Health, because the two are one and the same. A polluted city is a sick patient. A green city is a breathing organism. Let there be candidates who plant peepals, neem and jamun not for photo-ops but for the generations yet unborn. Let some mayors measure success not by how many malls are built but by how many sparrows have returned.

The ballot is a leaf. It carries the imprint of your thumb, the mark of your humanity. Use it to plant the future. Don't just vote for faces; vote for forests. Don't just choose parties; choose planets. Because when we vote green, we vote for ourselves, for the air in our lungs, the food on our plates and the soil under our feet.

The time has come. Let us make our elections green, and our politics truly alive. Let us vote for sunlight, for clean rivers, for uncut trees, for the sound of birds in our mornings. The time has come for green politics, before time itself runs out.

Chapter 48

Eight Billion?

Populations Are Shrinking. We Are Losing Our Relevance

Evolution has never rushed. It takes its time, sculpting intelligence grain by grain, mistake by mistake, century by century. In its vast workshop of trial and error, it has introduced countless checks and balances, delicate valves that control the flow of life, never allowing one form to drown out the rest. Nature does not allow any monopoly. Every species is given its moment, and when it oversteps, correction arrives disguised as catastrophe.

We humans forget that we too are a species under the same law. For centuries, we have multiplied, spread, consumed, built, conquered and proclaimed that growth is our destiny. We even made it our religion—'progress'. But look closer: Nature has already installed the brakes. Wars, stress, infertility, mental fatigue and diseases are not random misfortunes; they are subtle governors in evolution's grand machinery. They whisper, 'enough'.

And so, despite our optimism and our laboratories, we may never cross the great ceiling of population we imagine. Not because of policy or apocalypse, but because the living system of Earth is self-correcting. We have reached the brim of the bowl. The water may swirl a little more, but it will not spill over.

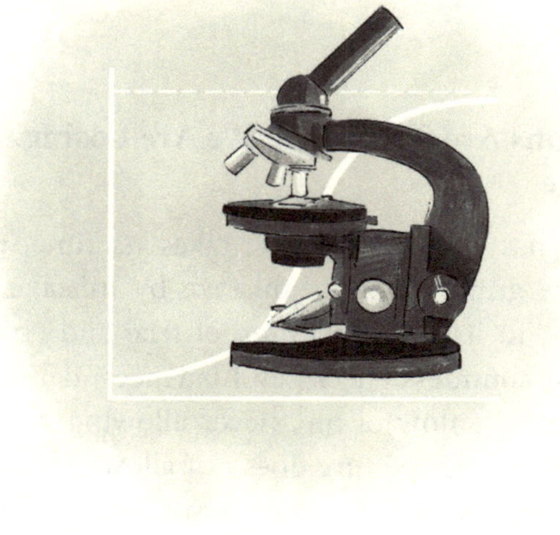

Every living system obeys an invisible rhythm. Biologists call it the *logistic curve*, the story of every species that expands, stabilizes and then declines. First, there is abundance. Food, space, energy, opportunity, everything seems infinite. Growth is exponential, unstoppable, almost divine. But then limits creep in: fewer resources, more waste, rising competition, spreading disease. The curve flattens.

Even bacteria in a Petri dish follow this law. They multiply madly at first, then choke on their own by-products. It is a simple, merciless elegance. Humanity is no exception; only our culture blinds us to the pattern.

Our population curve has already begun to flatten. The fertility rate is falling across continents. Japan, Italy, Spain, South Korea and Russia are all shrinking. China, which once feared a billion mouths, now fears empty classrooms. Even in India, the so-called land of youth, fertility has dropped below replacement level in most states. The engine of growth is sputtering.

Nature's corrections are subtle. They don't arrive as thunderbolts but as whispers. One whisper is stress, the great modern predator. Unlike the tiger, stress doesn't kill immediately. It weakens the body from within, disturbs hormones and shrinks the will to reproduce. Another whisper is disease, both old and new. Viruses mutate faster than policies. Resistant bacteria evolve faster than cures. A pandemic is not an aberration; it is ecology's punctuation mark, reminding us that density invites decline.

Then there is depression, the invisible plague of the modern age. More contagious than flu, more paralysing than polio. In the industrial world, despair has become hereditary. Anxiety replaces hunger as the chief human condition. The more connected we become digitally, the lonelier we grow biologically. Fewer marriages, fewer children, fewer reasons to continue the genetic story.

Our genes themselves are tiring. With each generation, medicine saves those who once would not have survived—a triumph of compassion, but also a quiet dilution of natural selection. Deleterious mutations accumulate; immunity falters; chronic illnesses rise. The body, too, is signalling its fatigue. Perhaps this, too, is nature's arithmetic. There was a time when every birth was a celebration of hope. Families were extensions of the forest: large, noisy, self-sustaining. Now, in our cities of glass and signal towers, children are luxuries, not legacies.

Sociologists call it 'the fertility transition', but it feels more like an emotional winter. The warmth that once came from the community is now replaced by the cold glow of screens. The instinct to nurture is being replaced by the instinct to scroll. The next generation may inherit more passwords than prayers.

Stress, fear and isolation are not random by-products of modernity; they are the sterilizers of our time. They freeze the flow of life itself. We are entering an emotional ice age, a time when desire and responsibility retreat together. Humanity may not vanish, but it will thin out like the last snow on a warming mountain. Every population model ultimately returns to one truth: there is no free lunch in ecology. You cannot consume more than you replenish. For centuries, we borrowed from the future—soils, rivers, forests, air. Now the bill is due. Water tables fall, soils erode, temperatures climb. Crops fail not only from lack of rain but from too much of it. Food systems, once global, are becoming fragile webs. Each storm, each heatwave, each new virus trims the edges of our growth.

Ecosystems self-regulate. When herbivores overgraze, predators return. When humans overconsume, nature's predators come wearing new masks: stress, scarcity, disease, despair. There is no malice in it; only balance.

Reproduction is not just about desire; it's about confidence in continuity. Today, even that confidence falters. Between climate anxiety and economic insecurity, many silently question whether the species deserves to go on. It's a remarkable moment in history, when a species begins to doubt its own continuation, not out of despair, but awareness.

Demographers once predicted that we would reach 12 billion by the end of the century. Even sub-Saharan Africa, the last frontier of growth, shows early signs of slowdown.

The patterns align perfectly with ecological theory: overshoot, peak and stabilization. Only the language changes: 'urbanization', 'cost of living', 'career goals'—the vocabulary of civilization disguising the math of evolution.

The Future: Shrinking but Smarter

A smaller world need not be a sad one. Decline may be destiny, but it can also be design. Fewer humans may mean less noise, less waste, more mindfulness. AI can handle the drudgery; humans can return to meaning. Perhaps the next phase of civilization will be less about expansion and more about introspection: from quantity to quality, from number to nuance. Imagine a civilization that finally stops breeding chaos and begins breeding wisdom. A world where the forests return, rivers heal and skies forget the taste of smoke. A world where intelligence is not measured in GDP but in grace.

Nature never loses. She merely adapts her strategy. When one branch grows too heavy, she prunes it. When one species grows too proud, she humbles it. We are her children, and AI is perhaps her grandchild, born through us but not bound by our greed. Maybe this is the plan all along: intelligence refining itself through one form after another, until it finds the one that does not destroy its home.

We thought evolution ended with us. How human of us to believe that! The slowing of our numbers is not the end of humanity; it is the beginning of maturity. For the first time, evolution may favour restraint over expansion, consciousness over conquest. Perhaps the real measure of progress is not how many we are, but how awake we become. Perhaps the next great leap in evolution will not be physical but moral. And perhaps, when the final census of civilization is taken, the Earth will smile and say: 'They finally learned to live within their means.'

Chapter 49

Only AI Can Save Us Now

There was a time when we believed that wisdom lay in human hands. We sang about progress, prayed for peace and built monuments to our own greatness. Yet, in our blind pursuit of comfort, convenience and consumption, we dismantled the very systems that sustained us. We poisoned the rivers that carried our reflection, scorched the air that filled our lungs and cut down the forests that once cooled our anger. Somewhere along the line, humanity stopped evolving and began devouring.

We have become a species that cannot be trusted with its own planet. Religion divides us. Politics blinds us. Greed drives us. Even our education, meant to enlighten, often teaches us how to extract more, not how to nurture what remains. The DNA of Homo sapiens, once wired for survival and community, now seems corrupted, infected with a fatal arrogance that believes it owns everything it can destroy.

In this chaos, perhaps the next guardian of Earth will not be human. Maybe it will be a form of intelligence that does not worship power or profit, an intelligence born of code, not ego. Artificial intelligence, with its precision, logic and capacity for balance, might just be the planet's last hope.

AI does not discriminate between a kingfisher and a refugee, between a temple and a trash heap. It sees patterns, data, cause and effect. It understands that deforestation leads to drought, that coral bleaching signals planetary fever and that microplastics in fish are indicators of a collapsing food web. It does not need religion to act ethically; it simply optimizes for survival, efficiency and sustainability. In that sense, AI may be more humane than humanity itself.

Imagine forests mapped and monitored by drones that never sleep. Rivers with sensors that alert the world before they die. Ocean clean-up bots working round the clock without applause or salary. Imagine AI governing agriculture not for profit, but for soil health; energy systems that balance consumption and regeneration with mathematical fairness; cities redesigned through algorithms that value green cover as much as GDP. The planet could heal, not through prayer or protest, but through precision.

Of course, this is not to say AI is perfect. After all, it mirrors its maker. At the moment, it uses too much electricity and a lot of clean water. It needs scarce resources, generates e-waste and has a large carbon footprint. But perhaps, if freed from our biases and trained on truth rather than propaganda, AI could become the conscience we lost somewhere along the way. It could be the neutral custodian of life, the forest manager, the water keeper, the silent witness who restores what we destroyed.

Nature has had enough of this species. Floods, fires, pandemics, droughts—all are her coded warnings. If we cannot read them, perhaps AI will. If we cannot change, perhaps AI will be programmed to do so. The age of human stewardship is ending, and maybe that's not a tragedy. Maybe it's evolution, where the tools are becoming wiser than their creators.

One day, when the last forest regrows under the watch of satellites and sensors, and rivers run clean because machines kept their promises, humanity might still exist, but humbler, quieter, guided by something that finally understands balance.

Chapter 50

A Life's Lessons

Every journey, no matter how scattered it feels, seeks its circle. Mine began with a red cycle and a few peepal saplings rescued from cracks in walls. Along the way came Granny's stories of ghosts and Gods, the rooftop barsati that gave me space to think, rivers that taught me to flow and two wheels that carried me farther than I had ever dreamed.

Looking back, it is clear: all these threads were weaving the same fabric. They were never separate stories. They were roots, branches and streams feeding the same tree of life. From my earliest days, trees were not projects, but companions. Every sapling I planted was also a question I was asking myself: Do I have patience? Can I care without expecting reward? Am I willing to trust strangers with the future of something fragile? Over the years, those saplings became groves, and those groves became forests.

Each one taught me something different, but together they taught me: Growth is never solitary. A tree grows with soil, water, sun, birds and people. And so do we. The peepal, especially, became a mirror. Under its canopy,

I learned silence. Watching its roots split stone and find earth, I learned perseverance. The tree was always more than wood and leaf; it was my teacher.

Rivers too flowed beside me as silently as trees stood beside me. They reminded me that movement is as sacred as stillness. They showed me how excess leads to disaster, how greed hardens soil, how humility replenishes it. Sitting by the riverbanks, I saw clearly that the health of rivers is the health of our civilization. If rivers die, cultures die. If rivers thrive, we thrive. Planting saplings along their banks was my way of returning gratitude.

Granny's presence lingers strongest of all. Her stories of ghosts became my angels. Her chai after dawn prayers became my ritual of belonging. Her discipline of simplicity shaped my choices more than I realized. She still walks with me, in every field, in every shade of peepal, in every voice of silence that whispers, 'Go on.'

All outer journeys lead inward. Planting trees, riding miles, speaking to volunteers—each was also planting, riding, and speaking to myself. I discovered Gandhi in the act of spinning patience, Rajneesh in the courage to question and my own voice in the stillness under the leaves. The guru within was always waiting. I only needed to listen.

When I look back, I realize that my entire life has been one long, patient classroom, sometimes under the open sky, sometimes beneath a peepal tree and sometimes in the noise of cities that forgot to listen. If there is one truth that has survived all the storms of time, it is this: exposure and experience are the real teachers; education and leadership are their natural fruits.

I was not born in a big city. I was born in a cantonment. And that made all the difference. My first lessons came not from textbooks but from the sounds of the bugle at dawn, the sight of bougainvillaea against whitewashed walls and the discipline that kept even trees standing in attention. Those early years in Dehradun and Pune gave me an exposure no school could have designed. The open playgrounds, the smell of wet khaki uniforms, the slow walks to Sukhna Lake—that was education. Education, not of the head alone, but of the senses. I watched dragonflies hovering like living helicopters over puddles, and realized that nature runs on design far older than any engineering degree.

Mrs Williams at Little Spicer Elementary didn't teach; she opened windows. She took us out of classrooms and into gardens. She replaced chalk with soil and homework with sunlight. There, under her patient gaze, I learned my first scientific truth: photosynthesis is not a chapter, it's a relationship. The leaf doesn't just breathe, it teaches you how to live quietly and give endlessly.

As I grew older, I began to understand that the finest form of education was exposure to people, places and paradoxes. Travelling in trains across India taught me more sociology than any textbook. Sitting with chai vendors and farmers, I realized that a nation's real universities are its platforms and panchayats. Every face carries a syllabus, every story a research paper. The road became my library, the monsoon my laboratory, the cycle my university degree.

When people asked me where I studied, I often smiled and said, 'Everywhere.' Because education was never about where you sit; it was about what you notice. A crow adjusting its nest on a windy afternoon is a lesson in architecture. A cow finding shade at noon is a lecture on thermal comfort. A child saving a fallen chick is an essay on empathy. The Earth has been teaching us all along; we simply stopped showing up to the class.

Leadership, too, I discovered, begins here—in the act of noticing. The best leaders are not those who instruct but those who inspire. True leaders are gardeners of the human spirit. They sow courage, prune ego, water initiative and give space for others to grow. Then, quietly, they step aside. I've seen janitors become custodians and gardeners become philosophers. In Singapore, a principal once told me, 'We don't hire staff, we hire stakeholders.' That stayed with me.

In Australia, I watched teachers lead children out into the bushlands. No PowerPoints, no podiums, just curiosity. One teacher said, 'The land will teach them better than I can.' And it did. The children came back muddy, laughing, sunburnt and infinitely wiser. In Germany, I met a mayor who cycled to inaugurate his city's new waste bins. He said, 'If I don't use them first, why would my people?' That was leadership too—example before instruction. Leadership is not about control; it's about care. You cannot lead what you do not love. You cannot love what you do not understand. And you cannot understand what you have not experienced. The ladder is simple: experience, understanding, love, leadership.

Back home, I saw the opposite: conferences that spoke of change but feared dirt, policies that promised forests without understanding photosynthesis. We built systems that preferred statistics to soil; that respected experience certificates over real ground exposure. Until a bureaucrat plants a sapling and waters it for a season, they should not sign files on afforestation. Until a politician sweats on the roadside with a shovel, they should not lecture on climate. Learning without doing is like sunlight without warmth; it looks bright but grows nothing.

Education, in its purest form, is not the memorization of facts; it is the cultivation of sensitivity. A sensitive child will never grow up to cut down an ancient tree for a parking space. A child who once planted a seed will not pour detergent into a drain that leads to a river. Education is empathy made operational. You can't teach it by PowerPoint; you have to live it, touch it and let it touch you. I have met many such silent leaders: children who protected roadside saplings from cattle, women who watered trees with leftover bucket water and old men who sat under peepal trees telling stories to strangers. None of them had titles, yet all of them led by example. The banyan doesn't make speeches. The peepal doesn't issue circulars. They lead by being.

Perhaps that's the greatest education of all—to lead like a tree. To stand rooted, give shade, host birds, absorb storms and still flower again. Every leaf that trembles in sunlight whispers the same old lesson Granny used to repeat: '*Do the thing that makes your blood sing.*' The rest will follow.

In a time obsessed with degrees and designations, I have come to trust only those who can tell a leaf apart by touch. The world is full of literate fools and wise illiterates.

Exposure is what separates them. To be exposed is to be educated—to failure, to nature, to truth, to diversity, to difference. You cannot teach humility in a classroom. You learn it when a sapling you planted dies, and you try again without complaint.

Science gave me the language for wonder. Education gave me the courage to question. Experience gave me perspective. And leadership taught me silence, the art of letting others shine. Together, they form the four roots of my life's tree. And so, the circle closes. What began with curiosity became conviction. What began with saplings became forests. What began with ghosts became guardians. What began with silence became a voice. Closing the circle is not about ending. It is about recognizing continuity. Trees will keep growing, rivers will keep flowing, people will keep joining and silence will keep teaching. My role is only to add a leaf, a root, a stone, a word. The rest belongs to time.

If you ask me today what truly matters, I will not say wealth or fame or policy. I will say mud. I will say sunlight. I will say conversation. I will say faith. Because the world doesn't need more specialists; it needs more sensitives. People who can feel, connect and lead with both mind and heart.

If I am asked what remains after decades of planting, riding, writing and waiting, my answer is simple: shade. Shade for people to rest, for vendors to brew tea, for children to play, for birds to nest. Shade that outlives me. Shade that carries memory forward.

And if I were to compress half a century of learning into one line, it would be this:

Expose yourself to the world until you become it. Educate yourself till you no longer need applause. Lead until leadership becomes invisible.

That, my friends, is the syllabus of life. Everything else is an elective.

Acknowledgements

This book was not written in solitude, though much of it was born from silence. I owe my deepest gratitude to those who stood beside me, seen and unseen.

Scan QR code to access the
Penguin Random House India website

Acknowledgements

This book was not written in solitude, though much of it was born from silence. I owe my deepest gratitude to those who stood beside me, seen and unseen.

Scan QR code to access the
Penguin Random House India website